Structure-activity correlations

and solid-state kinetic investigations of

iron oxide-based catalysts supported on SBA-15

vorgelegt von

M.Sc.

Nina Sharmen Genz

geb. in Dearborn

von der Fakultät II – Mathematik und Naturwissenschaften

der Technischen Universität Berlin

zur Erlangung des akademischen Grades

Doktor der Naturwissenschaften

Dr. rer. nat.

genehmigte Dissertation

Promotionsausschuss:

Vorsitzende: Prof. Dr. rer. nat. Maria Andrea Mroginski

Berichter/Gutachter: Prof. Dr. rer. nat. Thorsten Ressler

Berichter/Gutachter: Prof. Dr. rer. nat. Malte Behrens

Tag der wissenschaftlichen Aussprache: 11.03.2019

Berlin, 2019

Bibliographic information published by the Deutsche Nationalbibliothek

The Deutsche Nationalbibliothek lists this publication in the Deutsche
Nationalbibliografie; detailed bibliographic data are available
on the Internet at http://dnb.d-nb.de .

Zugl.: Berlin, Technische Universität, Diss., 2019

ISBN 978-3-8325-4924-4

Logos Verlag Berlin GmbH
Comeniushof, Gubener Str. 47,
10243 Berlin
Tel.: +49 (0)30 42 85 10 90
Fax: +49 (0)30 42 85 10 92
INTERNET: https://www.logos-verlag.de

"What is written without effort
is in general read without pleasure."
Samuel Johnson (1709-1784)

Zusammenfassung

Eisenoxide geträgert auf SBA-15 erwiesen sich als geeignete Modellkatalysatoren zur Untersuchung von Struktur-Aktivitätsbeziehungen während der selektiven Oxidation von Propen. Der Einfluss der Eisenbeladung, des Fe(III)-Präkursors, der Höhe der Pulverschüttung während der Kalzinierung, sowie der Molybdänzugabe auf die Fe_xO_y/SBA-15-Katalysatoren wurde untersucht. Die Eisenbeladung variierte zwischen 2.0 und 10.7gew.% und es wurden zwei Fe(III)-Präkursoren ((NH$_4$, Fe(III))-citrat und Fe(III)-nitrat Nonahydrat) verwendet. Die Kalzinierung erfolgte in einer dünnen (0.3 cm) oder dicken (1.3 cm) Pulverschüttung. Neben einer detaillierten Charakterisierung wurde die Anwendbarkeit von festkörperkinetischen Analysemethoden auf geträgerte Eisenoxide gezeigt. Struktur-Aktivitätsbeziehungen von Fe_xO_y/SBA-15-Katalysatoren ermöglichten die Bestimmung verschiedenster Syntheseparameter, welche die katalytische Aktivität beeinflussen.

Der Fe(III)-Präkursor beeinflusste die resultierenden Spezies, deren Größe, sowie strukturelle, festkörperkinetische und katalytische Eigenschaften. Der (NH$_4$, Fe(III))-citrat-Präkursor bewirkte kleinere, feiner verteilte Fe_xO_y-Spezies auf SBA-15. Dies ging einher mit einem höheren Maß an Mikroporenbefüllung, einer glatteren SBA-15-Oberfläche, sowie einem höheren Anteil an β-FeOOH-ähnlichen Spezies. Diese Unterschiede wurden dem unterschiedlich starken Chelateffekt der beiden Präkursoren, und ferner dem chemischen Gedächtnis der Fe_xO_y/SBA-15-Katalysatoren zugeschrieben. Außerdem waren sowohl die scheinbare Aktivierungsenergie der Reduktion als auch die Reaktionsrate und die Acroleinselektivität präkursorabhängig.

Die höhere Pulverschüttung während der Kalzinierung verlängerte die Aufenthaltsdauer der gasförmigen Zersetzungsprodukte des verwendeten Präkursors im Pulverbett. Infolgedessen erhöhte sich das Fe_xO_y-Spezieswachstum. Der Einfluss der variierten Kalzinierung auf die strukturellen, festkörperkinetischen und katalytischen Eigenschaften der Fe_xO_y/SBA-15-Katalysatoren war präkursor- und beladungsabhängig. Bei Verwendung des Fe(III)-nitrat Nonahydrat-Präkursors und bei geringen Eisenbeladungen war der Einfluss auf die resultierenden Fe_xO_y-Spezies stärker ausgeprägt.

Die gesteigerte katalytische Aktivität der Fe_xO_y/SBA-15-Katalysatoren infolge der Molybdänzugabe (Mo/Fe = 0.07/1.0–0.57/1.0) resultierte aus dem Dispersionseffekt und dem elektronischen Effekt des Molybdäns auf die Fe_xO_y-Spezies. Der Dispersionseffekt bewirkte kleinere Fe_xO_y-Spezies, während der elektronische Effekt deren Fe-O-Bindungen stärkte und somit die Reduzierbarkeit hemmte.

Abstract

Iron oxidic species supported on SBA-15 were introduced as model catalysts for deducing structure-activity correlations in selective oxidation of propene. Influence of iron loading, Fe(III) precursor, powder layer thickness during calcination, and molybdenum addition on Fe_xO_y/SBA-15 catalysts was investigated. Iron loading ranged from 2.0 through 10.7wt% and two Fe(III) precursors ((NH_4, Fe(III)) citrate and Fe(III) nitrate nonahydrate) were used. Calcination was performed either in thin (0.3 cm) or thick (1.3 cm) powder layer. Fe_xO_y/SBA-15 catalysts were characterized *ex situ* and *in situ* by a multitude of analyzing methods. Additionally, applicability of solid-state kinetic analysis methods was shown for supported iron oxidic species. Structure-activity correlations of Fe_xO_y/SBA-15 catalysts afforded determining various synthesis parameters accounting for the catalytic performance.

The Fe(III) precursor affected resulting species size, surface and porosity characteristics, and types of iron oxidic species. Additionally, solid-state kinetic properties and catalytic performance of resulting Fe_xO_y species were affected by the Fe(III) precursor. The (NH_4, Fe(III)) citrate precursor induced smaller, higher dispersed Fe_xO_y species on SBA-15. This was accompanied by a higher degree of micropore filling, a smoother SBA-15 surface, and higher amounts of β-FeOOH-like species. Precursor-dependent differences in structural characteristics were ascribed to the different strengths of chelating effect of the two precursors, and further, to the chemical memory effect of Fe_xO_y/SBA-15 catalysts. Moreover, apparent activation energy in reduction, as well as reaction rate, and acrolein selectivity showed a precursor dependence.

Thick powder layer during calcination extended the retention time of gaseous decomposition products of the used precursor in the powder layer, yielding an enhanced Fe_xO_y species growth. The influence of powder layer thickness during calcination on structural and solid-state kinetic properties, as well as catalytic performance was dependent on precursor and iron loading. The Fe(III) nitrate nonahydrate precursor and low iron loadings induced a more pronounced influence of powder layer thickness during calcination on resulting Fe_xO_y species.

Enhanced catalytic performance of Fe_xO_y/SBA-15 catalysts due to molybdenum addition (Mo/Fe = 0.07/1.0–0.57/1.0) was ascribed to the dispersion and electronic effect of molybdenum on Fe_xO_y species. The dispersion effect of molybdenum induced smaller iron oxidic species on SBA-15. Furthermore, the electronic effect of molybdenum strengthened the Fe-O bonds of the Fe_xO_y species, yielding a hindered reducibility.

Contents

Abbreviations

AHM	ammonium heptamolybdate
AMCSD	American Mineralogist Crystal Structure Database
a.u.	arbitrary unit
BET	Brunauer, Emmett, Teller
BJH	Barrett, Joyner, Halenda
CMD	catalytic methane decomposition
CMK-3	ordered mesoporous carbon
CT	charge transfer
dim.	dimension
DR	diffuse reflectance
DTA	differential thermal analysis
DTG	differential thermogravimetry
e.g.	for example (Latin "exempli gratia")
eq.	equation
et al.	and others (Latin "et alii")
EXAFS	extended X-ray absorption fine structure
Fe_xO_y	iron oxidic species
Fe_xO_y/SBA-15	iron oxidic species supported on SBA-15
Fe_xO_y/SBA-15_Th	iron oxidic species supported on SBA-15, obtained from thick layer calcination
FID	flame ionization detector
FHH	Frenkel, Halsey, Hill
FT	Fourier transform
FTS	Fischer-Tropsch synthesis
FWHM	full width at half maximum
G	Gaussian function
GC	gas chromatography
i.e.	that is (Latin "id est")
IUPAC	International Union of Pure and Applied Chemistry
JMAK	Johnson, Mehl, Avrami, Kolmogorov
KM	Kubelka, Munk
LC	linear combination
LCXANES	linear combination X-ray absorption near edge structure analysis
LMCT	ligand to metal charge transfer
MBS	Mössbauer spectroscopy
MCM-41	mesoporous silica (Mobile Composition of Matter No. 41)
m/e	mass-charge ratio
MFI	Mordenite Framework Inverted zeolites
MLCT	metal to ligand charge transfer

Mo_xO_y	molybdenum oxidic species
$Mo_xO_y\_Fe_xO_y$/SBA-15	molybdenum and iron oxidic species supported on SBA-15
MS	mass spectrometry
Nd:YAG	neodymium-doped yttrium aluminum garnet
NIR	near infrared
ODS	oxidative desulfurization
OFW	Ozawa, Flynn, Wall
PEEK	polyether ether ketone
RWGS	reverse water gas shift
SBA-15	mesoporous silica (Santa Barbara Amorphous type material No. 15)
SADs	structure-directing agents
SAED	small-angle electron diffraction
SKM	Schuster, Kubelka, Munk
TA	thermal analysis
TCD	thermal conductivity detector
TEM	transmission electron microscopy
TEOS	tetraethyl orthosilicate
TG	thermogravimetry
TPR	temperature-programmed reduction
unc.	uncalcined
UV	ultraviolet
Vis	visible
Vol%	volume percent
wt%	weight percent
XAFS	X-ray absorption fine structure
XANES	X-ray absorption near edge structure
XAS	X-ray absorption spectroscopy
XRD	X-ray diffraction
XRF	X-ray fluorescence

1 Introduction

1.1 Motivation

Chemical industry constitutes the third largest economic sector in Germany with a sales volume of more than 184 billion Euros per year. In the European Union, the largest chemical industry is that of Germany. The globally largest market of the chemical industry is China, followed by the USA, and afterwards by Germany. [1–4] Approximately 90% of all chemical processes are catalyzed. [5, 6] About one quarter of worldwide produced chemicals and intermediates are based on selective oxidation reactions. [7, 8] Among those, selective oxidation of propene to acrolein and further to acrylic acid is of particular importance. Acrolein serves as both intermediate for the synthesis of acrylic acid and raw material for the production of methionine, a synthetic racemic amino acid. The importance of methionine as animal-feed additive is exemplified by the annual production goal of 730 000 tons aimed by Evonik Industries AG starting in 2019. [9] Acrylic acid and its esters represent important raw materials for the fine chemicals industry. These monomers, with their reactive α,β-unsaturated carboxyl group, undergo polymerization reactions yielding polymers with various applications. The polymers are used as adhesives, as binding agents in the paper industry, as superabsorbent, and moreover, in the production of paints. [10] Industrial synthesis of acrylic acid proceeds in a two-step process. The first step is the oxidation of propene to acrolein, which is further oxidized to acrylic acid in the second step. The catalysts employed for this process are complex multicomponent metal oxides. Examples for applied metals are bismuth, molybdenum, iron, vanadium, cobalt, nickel, and tungsten. [10, 11] With the current catalysts, selective oxidation of propene to acrolein proceeds at a propene conversion of more than 90% and yields a selectivity towards acrolein of 80-85%. [12] Total or close-to-total conversions are hardly ever attained. Furthermore, limitations in catalytic oxidation processes constitute the formation of uneconomic by-products and the use of expensive feedstocks. Consequently, chemical industry aims at increasing both conversions and selectivity towards desired products for enhancing the efficiency of catalyzed processes.

In the last years, especially the approaching exhaustion of petrol feedstock effected a change in energy policies. Not only in automotive industry, but also in chemical industry, replacing fossil feedstocks by so-called bio-based feedstocks is required. Particularly in view of future green chemistry, this change in feedstocks is indispensable. [6, 13, 14] This affects also the selective oxidation of propene. Propene is the second most demanded product in petrochemical industry. [15] Glycerol as possible new bio-based feedstock for synthesis of acrolein recently attracted attention in catalysis research. Glycerol is abundantly available and relatively inexpensive as a by-product from biodiesel production. It can be transformed to acrolein by catalytic dehydration. [13, 14] However, this bio-based synthesis route is still in its early stages and an enormous research effort is required for generating an industrially applicable process. Therefore, interest in further improvement of already well-investigated selective oxidation of propene to acrolein reappeared. An improvement in both selectivity towards acrolein and propene conversion, at best without generating by-products, would lead to a considerably increased cost efficiency. Moreover, selective oxidation of propene would become *greener*. Such an improvement might be realized by optimized or newly designed catalysts. However, catalysts applied in selective oxidation reactions are highly complex and seem to become more complex while becoming more efficient. Therefore, rational catalyst design constitutes a promising alternative to conventional trial-and-error approach in developing optimized catalysts. [16–18] Rational catalyst design implies an in-depth understanding of reliable structure-activity correlations. In the last years, advancements in *in situ* analyzing methods enabled gaining deeper insight into important variables determining catalytic performance. Although *in situ* investigations are indispensable for elucidating structure-activity correlations, certain functions cannot be assigned unambiguously to certain structural motifs or individual components of the catalysts. Accordingly, distinguishing between influence of structural variety and chemical complexity of the metal oxide catalysts on their catalytic performance is challenging. Hence, reducing the number of variables affecting the catalytic performance by reducing the complexity of the industrial catalysts, is a promising approach towards a deeper understanding of heterogeneous catalysts. Model catalysts possessing a reduced complexity allow to correlate certain structural motifs or chemical functions with their catalytic performance. Model catalysts possess either a structural invariance while chemical composition is varied, or a compositional invariance and a structural variety. [17, 18] Binary metal oxides can be used as rather simple model catalysts without entirely repealing multifunctionality of complex industrial catalysts.

In heterogeneous catalysis, catalytic reactions occur on the surface of the catalysts, and the surface structure differs significantly from that of the bulk. Hence, bulk compounds barely participate in catalytic reactions. [5, 17] This raises an analytical problem, since many of the frequently applied analytical methods yield information on average bulk compounds rather than on surface structures. A promising approach to avoid this analytical problem and to study structure-activity correlations of surface structures, is the generation of supported metal oxide catalysts. Hence, highly dispersed metal oxides on well-defined support materials with high surface area constitute advantageous model catalysts. *In situ* investigations of these model catalysts allow to readily deduce structure-activity correlations without considering bulk compound properties. [19–21] Despite recent advances in *in situ* analyzing methods, detailed characterization of very small and highly dispersed supported metal oxides remains challenging. Hence, additional analyzing methods, such as solid-state kinetic analysis as applied for bulk systems, could be helpful in corroborating structure-activity correlations of supported metal oxides, and hence, promote the rational design of improved catalysts.

While forcing the currently existing catalytic processes into *greener* ones, the search for alternative *greener* catalysts becomes more relevant. Iron is the second most abundant metal and constitutes 4.7wt% of the earth crust. [22] Therefore, iron-based catalysts constitute a promising alternative to more expensive metal catalysts. Moreover, various iron compounds are commercially available or easy to synthesize. The probably most important advantage in view of principles of green chemistry, is that iron is not only less hazardous, but also less toxic for human health and environment. [23–25] Furthermore, Fe^{2+}/Fe^{3+} pairs are often used as redox promotors to improve redox properties of selective oxidation catalysts. [17, 26, 27] Replacing less environmentally friendly and more expensive metals by iron without diminishing the catalytic performance, or at best by increasing the efficiency of the process, would be an advancement in catalysis research.

1.2   Outline of this work

The objective of this work was the synthesis as well as the structural and functional characterization of iron oxidic species supported on SBA-15 regarding their applicability as model catalysts for selective oxidation of propene. Therefore, the influence of Fe(III) precursor and powder layer thickness during calcination on structural characteristics of resulting iron oxidic species was investigated with respect to iron loading.

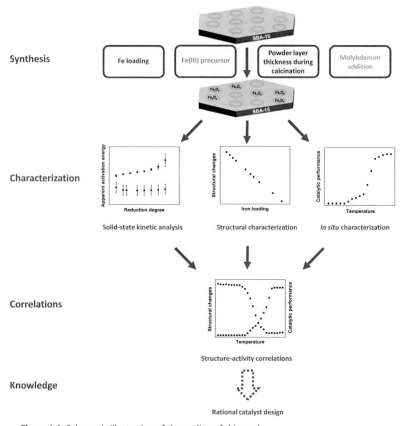

Figure 1.1: Schematic illustration of the outline of this work.

A multitude of *ex situ* analyzing methods was applied to ensure a detailed structural characterization of the model catalysts. Afterwards, investigated model catalyst system was *in situ* characterized. Solid-state kinetic analysis of the reduction was successfully applied to supported iron oxidic species. Thereby, additionally obtained information enlarged the knowledge of these model catalysts. Investigation of catalytic performance of Fe_xO_y/SBA-15 catalysts focused on the dependence on iron loading, precursor, and powder layer thickness during calcination. Variations in both solid-state kinetic properties and structural characteristics of supported iron oxidic species were subsequently correlated with their catalytic performance in selective oxidation of propene. Moreover, synthesis and characterization of iron and molybdenum mixed oxides supported on SBA-15 are presented in this work. A detailed *ex situ* characterization of these binary model catalysts was performed to elucidate the influence of molybdenum addition on structure of Fe_xO_y/SBA-15 catalysts. Additionally, molybdenum induced effects on catalytic performance and reducibility of supported iron oxidic species were studied. The outline of this work is schematically illustrated in Figure 1.1.

2 Scientific background

2.1 Supported metal oxides in catalysis

Supported metal oxides constitute a major group of heterogeneous catalysts. They are frequently used as selective oxidation catalysts or catalysts for other heterogeneously catalyzed reactions. Furthermore, supported metal oxides are often used as so-called environmental catalysts, which catalyze the transformation of undesired pollutants into non-noxious products. The support material is considered to be inert and commonly porous support materials are applied. [28, 29] Dispersing metal oxides on well-defined, high surface support materials yields in highly active and stable catalysts. These dispersed metal oxides are predominantly two-dimensional and differ significantly from bulk metal oxides. Depending on preparation method, metal oxide loading, nature of support, and interactions between metal oxides and support, several different metal oxide species can exist. At low metal oxide loadings, mainly isolated metal oxides result on the support. An increased loading induces the presence of oligomeric or polymeric metal oxide species, and even surface-bounded nano-crystalline metal oxides. Structural characterization of the active metal oxide species constitutes a mandatory starting point for deducing reliable structure-activity correlations of supported metal oxide catalysts. [28–30]

2.1.1 Support material

Mesoporous materials are used in a wide field of applications, including heterogeneous catalysis, sensor technology, gas storage, and adsorption. However, industrial applicability of mesoporous materials proves difficult, not least due to the crucial influence of the synthesis conditions on physiochemical properties of the resulting materials. [31, 32] In heterogeneous catalysis, mesoporous support materials are frequently used for enhancing the catalytic activity by dispersing metal oxides on these high surface support materials. Mesoporous materials possess specific surface areas between 600 and 1000 m^2/g and pore sizes ranging between 2 and 50 nm. A prominent representative of mesoporous materials is silica SBA-15, consisting of two-dimensional hexagonally ordered pore channels with a narrow pore size distribution. The mesopore channels of SBA-15 are connected by a network of micropores. [33–35]

Mesoporous materials are synthesized applying supramolecular aggregates as structure-directing agents (SDAs). [33, 34, 36, 37] For the synthesis of SBA-15, block copolymers, e.g. poly(ethylene oxide)-poly(propylene oxide)-poly(ethylene oxide) triblock copolymer, Pluronic P123, are commonly used. The advantages of using such an amphiphilic block copolymer as inorganic SDA are the biodegradability, commercial availability, low-cost, and non-toxicity compared to ionic or neutral surfactants as SDAs. The amphiphilic block copolymer consisting of hydrophilic head groups and hydrophobic tail groups, arranges in micelles when dissolved in polar solvents. Exceeding the critical micelle concentration, the block copolymer self-assembles into micellar liquid crystals, which in turn serve as template for the formation of a silica-based inorganic network around the organic aggregates after adding a silica source (e.g.

TEOS). Subsequently, the organic template is removed by calcination and a highly ordered mesoporous structure results. [33–38] Figure 2.1 illustrates the schematic representation of the synthesis of SBA-15.

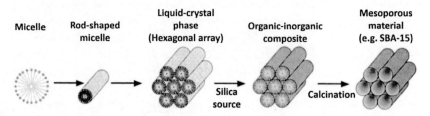

Figure 2.1: Schematic representation of the synthesis of mesoporous materials (adapted from [36]).

2.1.2 Supported iron oxides

Supported iron oxides are catalytically active in various reactions. They are, amongst others, catalysts for Fischer-Tropsch synthesis [39–44], Friedel-Crafts alkylations [45, 46], catalytic oxidative desulfurization (ODS) [47], catalytic methane decomposition (CMD) [25], NOx removal, and various selective oxidation reactions, such as selective oxidation of methane [48–50], benzene [51, 52], ethylbenzene [53], and propene [54]. Commonly, iron oxides are supported on ordered mesoporous silica materials, e.g. SBA-15 or MCM-41, Al_2O_3, ordered mesoporous carbon, e.g. CMK-3, zeolites, or MgO.

Due to the multitude of catalytic reactions based on iron containing catalysts, various synthesis procedures have been described in literature. However, it still remains challenging to control the iron species size and to achieve well-dispersed active iron species on the support. [55] Various authors reported the iron species being significantly influenced by the synthesis conditions, i.e. calcination temperature, precursors, promotors, and especially synthesis method. [25, 42, 48–50, 56] Al-Fatesh et al. investigated iron oxides supported on Al_2O_3 as catalysts for methane decomposition (CMD). [25] They reported a significant influence of calcination temperature on catalytic activity in CMD. The increased catalytic activity of Fe/Al_2O_3 samples calcined at lower temperatures (400-500 °C) was ascribed to a higher reducibility of the iron oxides. A correlation between catalytic activity in Fischer-Tropsch synthesis (FTS) and precursor, for iron oxides supported on α-Al_2O_3, was revealed by Torres-Galvis et al. [42]. Using an ammonium iron citrate precursor induced a more uniform distribution and an inferior aggregation of iron oxides on α-Al_2O_3, which further induced higher FTS activity and higher C_2-C_4 olefin selectivity. Lower catalytic activity and higher methane selectivity were attributed to higher aggregated iron oxides on α-Al_2O_3, due to the use of an iron nitrate precursor. The authors observed these differences in catalytic activity and product selectivity only for iron loadings of 5wt% and higher. Differences in catalytic activity in selective oxidation of methane were further reported as consequence of different synthesis procedures. [48–50] Arena et al. investigated the influence of varying synthesis procedures, i.e. adsorption-precipitation and incipient wetness method, on catalytic activity of low loaded FeO_x/SiO_2 samples. [50] They established a direct relationship between dispersion of iron species and catalytic activity in selective oxidation of methane. FeO_x/SiO_2 samples synthesized by adsorption-precipitation method showed an enhanced activity, ascribed to an improved dispersion of the active species. Besides, the active species were reported to be determined by their reducibility. Whereas poor reactivity was observed for mainly isolated iron species, and poor selectivity for Fe_2O_3 particles, the best catalytic activity resulted for two-dimensional $(FeO_x)_n$ oligomers, due to an optimal Fe-O bond strength. A correlation between higher dispersion of iron oxides on SiO_2 and higher catalytic activity in selective oxidation of methane was also reported by He et al. [49]. The higher dispersed iron species within FeO_x/SiO_2 samples prepared by sol-gel method induced a higher catalytic activity compared to those prepared by impregnation. They postulated the iron species at low iron loadings up to 0.5wt% synthesized by sol-gel method being mainly isolated in the SiO_2 matrix, whereas predominantly Fe_xO_y oligomers were ascribed to the samples synthesized by impregnation. Zhang et al. [48] reported similar to He et al. the isolated iron species, obtained at low iron loadings (< 0.1wt%), accounting best for selective oxidation of methane. The Fe_xO_y oligomers resulting at higher iron loadings showed a higher reducibility but a lower catalytic activity in selective oxidation of methane, and thus, an enhanced CO_2 formation. Tsoncheva et al. [56] investigated iron and copper oxide modified SBA-15 catalysts for methanol decomposition. A direct correlation between decreasing copper particle size and increasing catalytic activity was deduced. However, for iron oxides, they reported a complex correlation between simultaneous effects of iron particle size, reduced phase transformation, interactions between iron and the support, and catalytic activity. Despite various investigations of supported iron oxides in catalysis, structure-activity correlations are few understood, and no consensus exists concerning the active iron phase. Apparently, the resulting iron oxides on support materials are strongly dependent on synthesis conditions. Different active iron phases seem to be crucial for different catalytic reactions.

2.1.3 Supported iron and molybdenum mixed oxides

Previously, investigations of iron and molybdenum mixed oxide catalysts were reported aiming at understanding both interactions between iron and molybdenum and influence on catalytic activity. [57–62] Zhang et al. [57] investigated structural properties and catalytic performance in selective reduction of NO with NH_3 of Fe-Mo-SBA-15 catalysts. They ascribed the higher catalytic activity of bimetallic Fe-Mo-SBA-15 catalysts compared to corresponding monometallic catalysts to a decreased reducibility and an increased surface acidity due to molybdenum addition. Similar observations were reported by Ma et al. for Mo-Fe catalysts supported on activated carbon for Fischer-Tropsch synthesis. [58] Molybdenum addition resulted on the one hand in a suppressed reducibility of the iron species due to strong interactions between iron and molybdenum, and on the other hand in an increased iron dispersion and decreased iron species size. However, at 12wt% iron loading, the catalytic activity was heavily decreased due to a significant inhibition of reduction of the iron species. Furthermore, an increased catalytic activity in reverse water gas shift reaction (RWGS) was reported for Fe-Mo/Al$_2$O$_3$ catalysts compared to Fe/ Al$_2$O$_3$ catalysts. Kharaji et al. [59] revealed Fe-Mo/Al$_2$O$_3$ catalysts as novel RWGS catalyst with high CO yield and almost no by-product formation. Again, molybdenum addition resulted in strong interactions between iron and molybdenum, and hence, increased dispersion and hindered aggregation of the active iron species. The Fe-O bond was detected to be strengthened inhibiting the reducibility, and thus, the deactivation of the catalyst. However, the influence of molybdenum addition on supported iron oxides as selective oxidation catalysts is barely investigated. The effect of both dispersion and electronic effect of molybdenum on catalytic activity of supported iron oxides as selective oxidation catalyst remained largely unknown.

2.2 Selective oxidation reactions

Heterogeneously catalyzed selective oxidation reactions, using metal oxide catalysts, are commonly described by the so-called Mars-van-Krevelen [63] or redox-type mechanism (Figure 2.2). The reaction is assumed to proceed as sequence of consecutive steps, where oxidation of the reactant proceeding via an active site of the catalyst is followed by re-oxidation of the catalyst by gas-phase oxygen. According to this mechanism, the reactant in partial oxidation of an olefin, e.g. propene, is initially adsorbed on the catalyst surface. The subsequent activation of the adsorbed reactant by abstraction of a hydrogen in α-position to the double bond results in formation of an allylic intermediate. Afterwards, the reactant is oxidized by nucleophilic lattice oxygen of the catalyst and the reaction product is desorbed. Thereby, the catalyst is reduced. As last step in the catalytic cycle, the catalyst is re-oxidized by gas-phase oxygen, which requires an activation of molecular gas-phase oxygen. Only if conversion of gas-phase O$_2$ to nucleophilic lattice O^{2-} proceeds quickly, formation of selective oxidation products can be successful. Otherwise, catalyst re-oxidation might lead to the formation of electrophilic oxygen species (O$_2^-$, O$^-$, or O$_2$) on the surface of the catalyst, which attack the reactant at electron-rich positions, e.g. double bond. Subsequently, total oxidation and C-C bond cleavages are induced. Generally, lattice oxygen O^{2-} is considered to be responsible for the selectivity in selective oxidation reactions. [7, 12, 27, 30, 64]

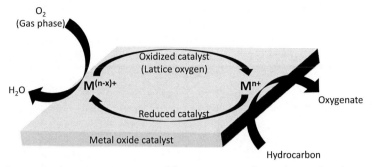

Figure 2.2: Schematic representation of the Mars-van-Krevelen mechanism for selective oxidation reactions catalyzed by metal oxides.

However, the detailed surface mechanism of heterogeneously catalyzed selective oxidation reactions is still discussed. Grasselli [7] proposed at least seven fundamental principles to be crucial for selective oxidation reactions catalyzed by metal oxides. Besides the importance of lattice oxygen, the role of metal-oxygen bond strength, host structure, site isolation, multi-functionality of active sites, cooperation of different phases, and redox properties are regarded. Currently applied catalysts consist amongst others of transition metal oxides with high redox activity, intermediate metal-oxygen bond strength, and not completely isolated metal ions. At least dimeric metal oxide species are postulated to be required for the complex transformations during the redox mechanism. Furthermore, only intermediate metal-oxide bond strengths are assumed to yield selective oxidation products. Too weak metal-oxide bond strengths, and therefore, too high oxygen availability will mainly result in undesired total oxidation products. A determining role is also ascribed to the redox properties of metal oxides. Only metals being able to perform a dynamic change in oxidation state are favored. Therefore, redox promotors, such as Fe^{2+}/Fe^{3+}, are frequently applied to improve redox properties of selective oxidation catalysts. [7, 27, 30]

2.3 Supported iron oxides in selective oxidation reactions

Supported iron oxides are frequently applied in selective oxidation reactions. However, most of the studies focus on particle size and dispersion effect rather than determining active iron sites or structure-activity correlations. One of the most prominent selective oxidation reaction catalyzed by supported iron oxides is the selective oxidation of methane to formaldehyde. [48–50, 65–67] Various reported suggestions concerning the active iron sites agree that the reducibility of the iron species accounts for the selectivity. However, there is not yet a consensus on whether isolated Fe^{3+} species or small Fe_xO_y oligomers possess the optimal reducibility. [48–50, 66, 67] Isolated, or mononuclear iron sites, interacting strongly with the support, lead to an inhibited oxygen availability, and thus, an increased selectivity towards the desired product is assumed. Conversely, Fe_2O_3 aggregates or nanoparticles possessing an enhanced reducibility and oxygen availability favor the total oxidation paths. These results comply with required site isolation and optimal metal-oxygen bond strength as proposed by Grasselli. [7] For selective oxidation of benzene to phenol catalyzed by Fe/MFI analogous results were reported. [52] Iron oxide nanoparticles yielded total oxidation products, whereas the highest phenol selectivity was obtained for the sample containing mononuclear iron sites. Wong et al. [53] investigated iron oxides supported on MCM-41 for selective oxidation of ethylbenzene. They determined the active iron species as distorted FeO surface species. Stabilization of such Fe(II) species was proposed to be enhanced due to Fe-O-Si linkages to the support. Although in selective oxidation reactions of olefins Fe^{2+}/Fe^{3+} pairs are applied as redox promotors for metal oxide catalysts, supported iron oxides alone are rarely investigated. Therefore, determining active iron species during selective oxidation of olefins, e.g. propene, constitutes a further advance in elucidating structure-activity correlations. Furthermore, fundamental understanding of the redox mechanism can be extended. Based on determined active iron sites, optimized, highly active, and selective metal oxide catalysts can be designed.

3 Characterization methods

3.1 Introduction

Identifying and characterizing the catalytically active species on the support material is one of the major challenges in heterogeneous catalysis. Hence, characterization of both supported iron oxide and iron and molybdenum mixed oxide catalysts requires a multitude of characterization methods. A complete characterization of the local structure of the supported iron oxide-based catalysts cannot be provided by applying only one individual characterization method. [20, 29] Using the various characterization methods summarized in Table 3.1, together with catalytic characterization and solid-state kinetic analysis, provides a powerful combination for deducing structure-activity correlations of supported iron and iron and molybdenum mixed oxide catalysts. In the following, a brief overview of basic principles of the applied characterization methods will be given. Additional information for a deeper understanding of these methods can be found in the cited literature.

Table 3.1 summarizes the applied characterization methods together with the provided information on oxidation state, coordination geometry, dispersion, magnetic properties, and quantitative determination of supported iron oxide-based catalysts and support material.

Table 3.1: Characterization methods together with the provided information on supported iron oxide-based catalysts and support material. N$_2$ physisorption, diffuse reflectance UV-Vis spectroscopy (DR-UV-Vis), X-ray absorption spectroscopy (XAS), powder X-ray diffraction (XRD), temperature-programmed reduction (TPR), thermal analysis (TA), Mössbauer spectroscopy (MBS), Raman spectroscopy (Raman), and X-ray fluorescence spectroscopy (XRF).

Characterization method	Supported Fe oxide-based catalysts					Support material
	Oxidation state	Coordi-nation	Dispersion	Magnetic properties	Quanti-tative	
N$_2$ physisorption	-	-	+/-	-	-	+
DR-UV-Vis	+	+	-	-	+/-	-
XAS	+	+	+/-	-	+/-	-
XRD	-	-	+/-	-	-	+
TPR	+/-	-	-	-	+	-
TA	-	-	-	-	-	+
MBS	+	+	+/-	+	+	-
Raman	+	+	+/-	-	-	-
XRF	-	-	-	-	+	+

3.2 Nitrogen physisorption

Nitrogen physisorption is one of the most important analytical methods for characterization of surface properties of supported catalysts. This method determines surface area and pore structure of supported systems. Physisorption denotes an interfacial phenomenon where a gas, the adsorptive, attaches to a solid surface, the adsorbent, by van der Waals interactions. A dynamic equilibrium establishes between adsorbent and adsorptive. The adsorption isotherm can be deduced from this dynamic equilibrium at isothermal conditions and expresses the amount of adsorbed gas, n, as function of the pressure, p/p_0:

$$n = f\left(\frac{p}{p_0}\right)$$ (3.1)

with equilibrium pressure, p, and saturation pressure of the adsorptive, p_0. [68] Physisorption isotherms are divided in six types by IUPAC. Figure 3.1 depicts the idealized types of isotherms associated with types of hysteresis loops proposed by IUPAC.

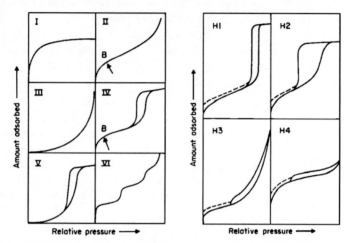

Figure 3.1: Left: Types of physisorption isotherms according to IUPAC. Right: Types of hysteresis loops according to IUPAC. [68]

The relevant type of physisorption isotherm in this work is type IV isotherm, indicative of mesoporous adsorbents. Type IV isotherms feature characteristic hysteresis loops being associated with capillary condensation in the mesopores of the adsorbent. The initial region of type IV isotherm resembles a type II isotherm where point B indicates the completion of monolayer adsorption and the beginning of multilayer adsorption. Type IV isotherms exhibit usually type H1 or H2 hysteresis loops. H1 hysteresis loops result from uniform, cylindrical pore channels with narrow pore radius distributions. Conversely, non-uniform pore channels and broad pore radius distributions effect H2 hysteresis loops.

Applying the method of Brunauer, Emmett, and Teller (BET) to physisorption isotherms enables determining specific surface area of mesoporous materials. According to BET method, the specific surface area, $a_{s,BET}$, is calculated considering the area covered by adsorptive, σ, and the capacity of the monolayer, n_m:

$$a_{s,BET} = n_m \cdot N_A \cdot \sigma$$ (3.2)

with Avogadro constant, N_A. [69]

In addition to the specific surface area, the pore size distribution of mesoporous materials can be determined using physisorption isotherms. The method of Barret, Joyner, and Halenda (BJH) is based on the phenomenon of capillary condensation and derived from the Kelvin equation. In this approach, theoretical emptying of the pores by stepwise reduction of the relative pressure, p/p_0, is considered by a thinning multilayer. Assuming cylindrical pores, the pore radius, r_P, is given by:

$$r_P = r_K + t \tag{3.3}$$

with Kelvin radius, r_K, and thickness of the multilayer, t. [70]

Using the adsorption data of the multilayer region results in information on roughness of the surface of mesoporous materials by determining the fractal dimension. Hence, the fractal dimension, D_f, is a measure for the roughness of the surface. An ideal structure with a smooth surface possesses a fractal dimension of 2. Conversely, a real rough surface possesses a fractal dimension of 3. Values of fractal dimension between 2 and 3 are indicative of materials with a fractal surface. Generally, Frenkel-Halsey-Hill method (FHH) is applied for determining the fractal dimension of porous materials. [71]

In case of a fractal surface, FHH method results in:

$$\ln\left(\frac{p_0}{p}\right) = k \cdot \left(\frac{n}{n_m}\right)^{\frac{s}{D_f - 3}} \tag{3.4}$$

with relative pressure, p/p_0, constants, k and s, amount of adsorbed gas, n, and monolayer capacity, n_m. This equation is solely valid with the assumption of negligible capillary condensation and surface tension. Otherwise, the modified FHH method has to be applied for determining the fractal dimension:

$$\frac{n}{n_m} = \left(\frac{\ln\left(\frac{p_0}{p}\right)}{k}\right)^{D_f - 3} . \tag{3.5}$$

From the slope of the resulting straight line of the log-log plot of the FHH or modified FHH method, the fractal dimension can be calculated. [71, 72]

3.3   Diffuse reflectance UV-Vis spectroscopy

Diffuse reflectance UV-Vis spectroscopy in ultraviolet (UV), and visible (Vis) region of the electromagnetic spectrum constitutes a major characterization method for heterogeneous catalysts. Irradiation of the sample with light induces electron transfer processes. In metal ions ligand-to-metal (LMCT) and metal-to-ligand (MLCT) charge transfer transitions, as well as by Laporte rules forbidden d-d transitions can occur. In contrast to UV-Vis spectroscopy of solutions and gas phases, and crystals, that of powder samples has to be performed in diffuse reflectance mode instead of transmission mode, which is denoted diffuse reflectance UV-Vis spectroscopy (DR-UV-Vis). DR-UV-Vis spectroscopy of heterogeneous catalysts, such as supported metal oxides, results in various structural information on the metal ions, e.g. oxidations state, coordination environment, local symmetry, and dispersion on the support. [73, 74] Irradiation of a powder sample with light is followed by a complex combination of diffuse reflection, total reflection, and single and multiple scattering processes. Schuster, Kubelka, and Munk developed a theory for describing the complex processes during DR-UV-Vis spectroscopy of powder samples. In the Schuster-Kubelka-Munk (SKM) theory, the sample is irradiated by light fluxes being perpendicular to the sample surface. Thereby, the incident and scattered light are approximated as two fluxes with opposing directions. [73, 75]

The Schuster-Kubelka-Munk function describes the relationship between experimental determined diffuse reflection of the sample, R_∞, and the absorption, K, and scattering, S, coefficient, accounting for light losses due to absorption and scattering, respectively:

$$F(R_\infty) = \frac{K}{S} = \frac{(1 - R_\infty)^2}{2R_\infty}.$$
(3.6)

The diffuse reflection of the sample, R_∞, is calculated by the quotient of experimentally determined reflection of the sample and the reflection of an ideal, non-absorbing white standard, e.g. MgO, $BaSO_4$, or Spectralon©. Applicability of the SKM theory is only given by fulfilling the following conditions:

- Completely diffuse monochromatic irradiation
- Isotropic light scattering
- Infinite layer thickness
- Low concentration of absorbing centers
- Uniform distribution of absorbing centers
- Absence of fluorescence.

Typically, experimental setup consists of an integration sphere, which is coated with a white standard (e.g. $BaSO_4$ or Spectralon©). Such an integration sphere collects the diffuse scattered light of the sample. Further details on DR-UV-Vis spectroscopy can be found in literature. [75, 76]

3.4   X-ray absorption spectroscopy

X-ray absorption spectroscopy (XAS) is a powerful element specific analytical method for elucidating electronic properties and local structure of the absorber atom. Especially the possibility of investigating amorphous materials and biological systems turns XAS into an indispensable and frequently applied method. The advantage of applicability of XAS to systems with no long-range ordering is particularly relevant for supported catalysts. [73, 77] The absorption of X-ray photons by the sample can lead to excitation of an electron from a core level. In the case of the incident X-ray photons reaching the binding energy of a core electron of the absorber atom, the electron from a core level is excited to an empty level above the Fermi level. This results in a sharp rise in absorption, denoting the absorption edge. Each element has a characteristic absorption edge depending on the energy of the core level.

In X-ray absorption spectra two regions can be discerned (Figure 3.2). The X-ray absorption near edge structure (XANES) and the extended X-ray absorption fine structure (EXAFS). [73, 77, 78]

Typically, the XANES region extends to 50 eV above the absorption edge and provides information on local electronic and geometric structure of the absorber atom. The spectral shape of the XANES region is dominated by multiple scattering and electron correlation effects. The XANES pre-edge region arises from the excitation of an electron from a core level to an unoccupied state close to the Fermi level. For these transitions, selection rules ($\Delta l = \pm 1$, $\Delta j = \pm 1$, $\Delta s = 0$) are valid which lead to the predominant transitions from s to p orbitals around the K-edge, and from p to d orbitals around the L_{II}- and L_{III}-edges. However, the final states of the transitions may also be hybridized orbitals effecting characteristic features in the pre-edge region due to coordination symmetry and oxidation state. [77]

Within the EXAFS region, scattering of the outgoing photoelectron by neighboring atoms is predominant. The outgoing photoelectron wave will be scattered back when reaching the neighboring atoms. Hence, the fine structure of the absorption coefficient results from interference between backscattered photoelectron wave and outgoing electron wave. The EXAFS function, $\chi(k)$, as function of wavenumber, k, is the sum of the scattering contributions of all neighboring atoms and can be used to describe the EXAFS modulations. [77]

$$\chi(k) = \sum_j A_j(k)\sin\left(2kR_j + \phi_j(k)\right) \tag{3.7}$$

with j referring to the jth coordination shell around the absorber atom, the amplitude, $A_j(k)$, i.e. scattering intensity due to the jth coordination shell, the distance, R_j, between absorber atom and atom in the jth coordination shell, and the total phase shift, $\phi_j(k)$. The EXAFS function assumes a single-scattering approximation. [73]

The amplitude, $A_j(k)$, of each scattering contribution is dependent on the number of neighboring atoms in a coordination shell and given by:

$$A_j(k) = \frac{N_j S_0^2(k) F_j(k)}{kR_j^2} \cdot e^{(-2k^2\sigma_j^2)} \cdot e^{(-\frac{2R_j}{\lambda(k)})} \tag{3.8}$$

where N_j is the number of neighboring atoms in the jth coordination shell, $S_0^2(k)$ is the correction for relaxation effects in the absorber atom, $F_j(k)$ is the backscattering amplitude in the jth coordination shell, σ^2 is the so-called Debye-Waller factor describing the mean-squared displacement of atoms in the sample, $\lambda(k)$ is the inelastic mean free path of the electron, and R_j is the distance between absorber atom and atom in the jth coordination shell.

Commonly, the EXAFS function is Fourier transformed, resulting in a pseudo radial distribution function. The $\chi(k)$ is usually multiplied with k^1 or k^3 to emphasize light or heavy atoms before Fourier transformation. Analysis of the Fourier transformed EXAFS function provides information on coordination environment of the absorber atom. [73, 77, 78]

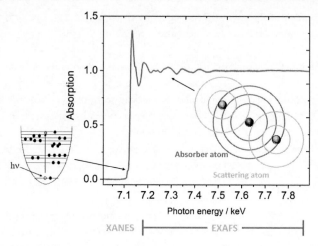

Figure 3.2: X-ray absorption spectrum with a schematic representation of the process at the absorption edge. XANES region: An electron from core level is excited to a higher unoccupied level. EXAFS region: The photoelectron wave (circles) emitted from the absorber atom (grey) is backscattered when reaching the nearby scattering atoms (green).

3.5 Powder X-ray diffraction

Powder X-ray diffraction (XRD) is one of most frequently applied analytical methods for determining long-range ordered structures of catalysts. This non-destructive technique is based on interactions between electromagnetic radiation and matter. The incoming monochromatic X-ray photons are elastically scattered by the electrons of atoms arranged in a periodic structure. The scattered X-ray photons interfere constructively or destructively depending on the distance between the lattice planes and the angle between lattice planes and incident X-ray photons. Detectable X-ray photons result only from constructive interference, being described by the Bragg equation. The Bragg equation describes the relationship between constructive interference of incident X-ray photons, angle, θ, between lattice planes and incident X-ray photons, and lattice spacing, d:

$$n\lambda = 2d \, sin\theta \tag{3.9}$$

with diffraction order, n, and wavelength of the X-ray photons, λ.

XRD investigations are used for determining crystalline phases, structural properties, and microstructures of the sample. Diffractograms are recorded as function of diffraction angle, 2θ. Based on the position of the diffraction peaks in the diffractogram the metric of the unit cell can be determined. Furthermore, qualitative phase composition can be identified. Analysis of intensity and shape of the diffraction peaks provides information on atomic positions, domain size, and micro-strain. [79, 80]

For mesoporous SBA-15 with a two-dimensional hexagonal crystal system, lattice parameter, a_0, can be calculated by inserting Miller indices, hk, in the Bragg equation:

$$a_0^2 = \frac{\lambda^2}{3 \cdot sin^2\theta}(h^2 + k^2 + hk) \qquad (3.10)$$

with wavelength, λ, and angle, θ.

Calculation of the wall thickness, d_w, of SBA-15 can be performed by using the lattice parameter, a_0, and the pore radius, r, from N_2 physisorption measurements in equation (3.3).

$$d_w = a_0 - 2r. \qquad (3.11)$$

3.6 Temperature-programmed reduction

Temperature-programmed reduction (TPR) constitutes an analytical method for investigating reduction processes of solid states with the involvement of gas phase components. The sample is exposed to a reducing gas mixture while temperature is increased with a constant heating rate. Typically, CO (CO-TPR) or H_2 (H_2-TPR) in argon or nitrogen are employed as reducing gas mixture. Hereafter, a short explanation of H_2-TPR will be given which applies analogously to CO-TPR. The sample is placed in a fixed-bed reactor and exposed to a H_2 gas mixture while the temperature is raised linearly. A thermal conductivity detector (TCD) is used for determining the gas composition at the outlet of the reactor, and thus, the rate of reduction is monitored. Calculating the total amount of consumed H_2 permits identifying reduction degree and moreover oxidation state of the sample. Besides, TPR experiments can be applied for kinetic analysis. The TPR profile depends on both sample properties and experimental parameters. Therefore, it is essential to ensure the TPR profile being only affected by sample properties. Experimental parameters like gas flow, H_2 concentration, amount of reducible species, and heating rate should be kept constant to ensure comparable measurements. [8, 81]

3.7 Solid-state kinetic analysis

Kinetic investigations constitute a method for analyzing reaction rates and observing chemical reactions as function of time. Hereby, reaction mechanism can be suggested, and reaction rates can be quantified dependent on different state variables. Heterogeneous kinetics in solid states, i.e. of reactions between solid states and gases, differ significantly from homogeneous kinetics in gas phases or liquids. Experimental measurements for solid-state kinetic analysis can be performed under either isothermal or non-isothermal conditions. Dependent on reaction conditions, fundamentally different analysis methods are required. Moreover, in contrast to isothermal conditions, solid-state kinetic investigations under non-isothermal conditions require a more complex mathematical analysis. In heterogeneous catalysis, evolution of structure and function of the catalysts are frequently determined under non-isothermal conditions. Hence, additional solid-state kinetic analysis of experimental data measured under these conditions may be helpful in corroborating structure-activity correlations. [82–86]

For solid-state kinetic analysis of data measured under non-isothermal conditions, two different approaches can be distinguished. First, solid-state kinetic data can be analyzed by model-independent methods, such as Kissinger method [87] or isoconversional method of Ozawa, Flynn, and Wall (OFW) [88, 89]. Whereas the Kissinger method yields one apparent activation energy of the rate-determining step, the OFW method yields an evolution of apparent activation energy as function of reaction degree, α. Model-independent kinetic analysis is not based on any model assumptions, consequently the "kinetic triple" (apparent activation energy, E_a, preexponential factor, A, of the Arrhenius-type temperature-dependence of the rate constant, and suitable solid-state reaction model, $g(\alpha)$) cannot be identified. Therefore, a second complementary approach to solid-state kinetic analysis is required. Model-dependent solid-state kinetic analysis employs several solid-state kinetic reaction models, $g(\alpha)$. After identifying the suitable solid-state reaction model, the "kinetic triple" can be determined. [86]

3.8 Catalytic characterization

Investigation of performance of a catalyst mainly aims at determining reactivity and selectivity of a reaction. Therefore, catalytic characterization of the catalysts can be performed using a fixed-bed reactor with connected gas chromatography system and mass spectrometer. After reactants passed through the catalyst bed, the reaction products at the outlet of the reactor can be analyzed by gas chromatography (GC) and mass spectrometry (MS).

Gas chromatography constitutes a qualitative and quantitative analytical method which is based on the separation of analyte mixtures, i.e. reaction product mixture. Separation of the components of the sample results from different interactions between sample in a mobile phase and stationary phase. A typical gas chromatography measurement starts with the injection of the sample in the inert gaseous mobile phase. The mobile phase carries the sample through the column, which serves as stationary phase. While passing through the column, the different components of the sample perform a dynamic process of adsorption and desorption between mobile and stationary phase. This dynamic process is strongly influenced by the chemical and physical properties of the sample. The stronger the interactions between analyte component and stationary phase, the higher the retention time of the component. As a result of the varying retention time, differences in migration velocity arise, which further lead to a separation of the different analyte components. The retention time defines the required time of a component for passing through the column from the point of injection to the point of detection. Connected detectors can analyze the different components at the outlet of the column as discrete signals. The detected signals are depicted as function of time in so-called chromatograms. Analyzing peak positions and peak areas enables a qualitative and quantitative analysis, respectively. [90–92] Further details on various detectors, reactors, and setups can be found elsewhere. [28, 90, 91, 93]

3.9 Thermal analysis

Thermal analysis comprises all analytical methods which detect physical and chemical properties of a substance or substance mixture as function of temperature or time, while the sample undergoes a controlled temperature program. [94] Investigations of thermal stability, decomposition or dehydration processes, oxidation reactions, or physical processes, such as evaporation and sublimation, are frequently conducted by thermal analysis. With thermogravimetry (TG) the mass of the sample is continuously monitored as function of temperature or time using a sensitive thermobalance. Commonly, the temperature is increased with a constant heating rate. Differential thermogravimetric curves (DTG) result from the first derivative of the TG signal. DTG curves are particularly applied to facilitate the identification of mass changes during overlapping reactions. [95, 96] More often than not, the thermobalance is coupled to a differential thermal analysis (DTA) in order to investigate reactions proceeding without mass changes. With DTA the temperature difference between sample and a thermally stable reference is determined, while both undergo a controlled linear temperature program. DTA measurements are only possible if the sample exchanges heat with the environment during a phase transformation or a chemical reaction. [95, 96]

3.10 Mössbauer spectroscopy

Mössbauer spectroscopy constitutes a nuclear technique providing information of oxidation state, lattice symmetry, and magnetic properties of solid states. This analytical method is limited to elements exhibiting the Mössbauer effect. The Mössbauer effect denotes the recoil-free nuclear resonance of γ-radiation. The resonant absorption of electromagnetic radiation (γ-photons) of the Mössbauer effect is similar to the resonance in acoustics and optics. The principle of nuclear resonance spectroscopy of γ-radiation is displayed in Figure 3.3.

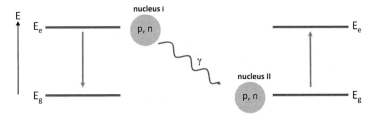

Figure 3.3: Principle of nuclear resonance spectroscopy of γ-radiation.

A nucleus I, with a proton number p and a neutron number n, is excited. The lifetime of this excited nucleus I is very short but measurable. During the transition from the excited energetic state E_e into the original energetic state E_g the difference in energy is emitted as γ-quantum. If this emitted γ-quantum subsequently hits an unexcited nucleus II with the same p and n in the energetic state E_g, an absorption follows and nucleus II transfers to the energetic state E_e. The energy difference between excited and original energetic state is crucial for this resonance process. The energy of the emitted γ-quantum has to be equal to the energy difference between the two considered energetic states. For achieving this, a recoil must be suppressed. In solutions or gas phases momentum conservation leads to a recoil of excited nuclei by emitting a γ-quantum. Conversely, excited nuclei in solid lattices cannot recoil as if they were free, and thus, the recoil energy is passed over the entire lattice. A recoil in solid lattices can be prevented by high mass and low temperatures. [77, 97, 98]

Mössbauer spectra show at least one resonance line. The number and position of the resonance lines are indicative for the interactions between nuclei and electromagnetic field. The electric monopole interaction between nuclei and electrons at the nucleus results in an isomer shift. Thus, the isomer shift yields information on oxidation state and binding properties. The electric quadrupole splitting arises from interactions of the electric quadrupole moment of the nucleus with an electric field gradient and is used for deducing information on ligand field effects and molecule symmetry. Conversely, magnetic hyperfine splitting results from interactions of the nuclear magnetic dipole moment and the magnetic field at the nucleus. Magnetic hyperfine splitting values are required for determining the magnetic properties of the sample. Further details about Mössbauer spectroscopy can be found in the literature. [77, 97, 98]

3.11 Raman spectroscopy

The Raman effect constitutes an inelastic scattering of electromagnetic radiation by matter. Similar to Infrared spectroscopy (IR), Raman spectroscopy also records transitions between different vibrational states. Whereas in IR spectroscopy, the absorption is measured as function of the frequency of the incident radiation, in Raman spectroscopy, the difference in frequency resulting from inelastic scattering in the sample is measured. In Raman spectroscopy, the sample is irradiated with a monochromatic laser beam. Afterwards, the excited sample can relax in two different ways. The majority of the excited molecules returns to its original energy state, thus an elastic scattering process (Rayleigh scattering) takes place. A minor amount of the excited molecules relaxes by emitting a different energy compared to the absorbed energy. This process is called inelastic scattering. The Raman scattering distinguishes two different inelastic scattering processes. The emitted energy of the excited molecules during relaxation is mostly lower than the absorbed energy (Stokes scattering). Thus, a decreased frequency is observed. Occasionally, an increased frequency due to a higher emitted energy compared to the absorbed energy is observed (Anti-Stokes scattering). [73, 99]

Raman scattering exhibits only a small cross-section representing a disadvantage of Raman spectroscopy. The Rayleigh band, resulting from elastic scattering of the incoming photons, is about three orders of magnitude stronger than the Stokes bands. For increasing the Raman signal, higher laser intensities are required. Due to a possible decomposition of surface species and a sample heating induced by high laser intensities, the use of such laser intensities is limited. Moreover, detection of weak Raman bands may be hindered by sample fluorescence.

Two major improvements concerning the difficulties in Raman spectroscopy can be achieved by using ultraviolet lasers as irradiation source. First, the Raman peak is shifted out of the visible spectral region where fluorescence occurs. Second, the cross-section, which is a function of the fourth power of the frequency, is increased. However, difficulty of compromising between minimal wavelength and maximal intensity of the laser still remains. [77]

3.12 X-ray fluorescence analysis

X-ray fluorescence analysis (XRF) constitutes a frequently used analytical method for qualitative and quantitative analysis of metal oxide catalysts. This method is based on irradiation of the sample with X-rays resulting in element specific fluorescence radiation. The characteristic fluorescence radiation can be used to identify the chemical composition of the sample. [92, 99]

Irradiation of the sample with X-ray photons results in removal of an electron from a core shell by the photoelectric effect. An electron of a higher energetic level fills the resulted core hole while emitting fluorescence radiation. The dependence of the energy of the emitted fluorescence radiation from both atomic number of the element, Z, and main quantum number, n, of the involved shells ($n_1 > n_2$) is described by Moseley's law:

$$h\nu = Rhc(Z - \sigma)^2 \left(\frac{1}{n_1} - \frac{1}{n_2}\right) \tag{3.12}$$

with Planck constant, h, frequency, ν, Rydberg constant, R, light velocity, c, and shielding constant, σ. Depending on the atomic number of the element, the fluorescence process competes with the Auger process. For lighter elements, the Auger process predominantly occurs. In this case, the energy released, when the electron returns to the ground state, is not emitted as fluorescence radiation, but transferred to another electron of a higher energetic level, which is consequently removed from the atom as Auger electron. [92] Figure 3.4 illustrates the electron excitation process by an X-ray photon that either yields the emission of an X-ray fluorescence photon or an Auger electron.

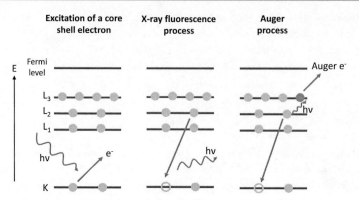

Figure 3.4: Schematic description of electron excitation process by an X-ray photon (left), relaxation process by emitting an X-ray fluorescence photon (middle), and relaxation process by emitting an Auger electron (right).

4   SBA-15 as suitable support material

4.1   Introduction

For investigating supported metal oxide catalysts, it is indispensable to start with a detailed characterization of the support material to ensure successful synthesis, and further, suitability of the as-prepared support material. In heterogeneous catalysis, suitable support materials have to fulfill the following requirements. High specific surface area, narrow pore size distribution, uniform pore structure, low catalytic activity, and thermal stability belong to the most important requirements. [28, 29] Hence, this chapter aims at characterizing the synthesized support material SBA-15 to ensure the successful synthesis and the suitability as support material for Fe_xO_y-based catalysts.

4.2   Experimental

4.2.1   Synthesis of SBA-15 support material

Mesoporous silica SBA-15 was prepared according to Zhao et al. [33, 34]. The surfactant, Pluronic® P123, was dissolved in a mixture of deionized water and HCl (37%) and the reaction mixture was stirred at 308 K for 24 h. Tetraethyl orthosilicate (TEOS) was added to the solution and the reaction mixture was stirred at 308 K for 24 h and then hydrothermally treated in pressure-resistant bottles at 388 K for 24 h. The obtained white solid was filtered, washed with a mixture of deionized water and ethanol (20:1), air-dried, and calcined. Calcination was carried out in three steps. (I) 378 K for 135 min, (II) 453 K for 3 h and (III) 873 K for 5 h. The heating rate was kept at 1 K/min.

4.2.2   Characterization of SBA-15 support material

Nitrogen physisorption

Nitrogen adsorption/desorption isotherms were measured at 77 K using a BELSORP Mini II (BEL Inc. Japan). Prior to measurements, the samples were pre-treated under reduced pressure (10^{-2} kPa) at 368 K for 35 min and kept under the same pressure at 448 K for 15 h (BELPREP II vac).

Powder X-ray diffraction

Powder X-ray diffraction patterns were obtained using an X'Pert PRO diffractometer (PANalytical, 40 kV, 40 mA) in theta/theta geometry equipped with a solid-state multi-channel detector (PIXel). Cu K_α radiation was used. Wide-angle diffraction scans were conducted in reflection mode. Small-angle diffraction patterns were measured in transmission mode from 0.4° through 6° 2θ in steps of 0.013° 2θ with a sampling time of 90 s/step.

Transmission electron microscopy

Transmission electron microscopy (TEM) images were recorded on a FEI Tecnai G² 20 S-TWIN microscope equipped with a LaB_6 cathode and a 1 k x 1 k CCD camera (GATAN MS794).

Acceleration voltage was set to 220 kV and samples were prepared on 300 mesh Cu grids with Holey carbon film. Measurements were conducted in collaboration with ZELMI (Zentraleinrichtung für Elektronenmikroskopie) at TU Berlin.

Catalytic characterization

Quantification of catalytic activity in selective oxidation of propene was performed using a laboratory fixed-bed reactor connected to an online gas chromatography system (CP-3800, Varian) and a mass spectrometer (Omnistar, Pfeiffer Vacuum). The fixed-bed reactor consisted of a SiO_2 tube (30 cm length, 9 mm inner diameter) placed vertically in a tube furnace. A P3 frit was centered in the SiO_2 tube in the isothermal zone, where the sample was placed. Overall sample masses were 0.25 g to achieve a constant volume in the reactor and to minimize thermal effects. For catalytic testing in selective oxidation of propene, a gas mixture of 5% propene and 5% oxygen in helium was used in a temperature range of 298-653 K with a heating rate of 5 K/min. Reactant gas flow rates of propene, oxygen, and helium were adjusted through separate mass flow controllers (Bronkhorst) to a total flow of 40 ml/min. All gas lines and valves were preheated to 473 K. Hydrocarbons and oxygenated reaction products were analyzed using a Carbowax 52CB capillary column, connected to an Al_2O_3/MAPD capillary column (25 m x 0.32 mm) or a fused silica restriction (25 m x 0.32 mm), each connected to a flame ionization detector (FID). Permanent gases (CO, CO_2, N_2, O_2) were separated and analyzed using a "Permanent Gas Analyzer" (CP-3800, Varian) with a Hayesep Q (2 m x 1/8'') and a Hayesep T packed column (0.5 m x 1/8'') as precolumns combined with a back flush. For separation, a Hayesep Q packed column (0.5 m x 1/8'') was connected via a molecular sieve (1.5 m x 1/8'') to a thermal conductivity detector (TCD). Additionally, product and reactant gas flow were continuously monitored by a connected mass spectrometer (Omnistar, Pfeiffer) in a multiple ion detection mode. Further details on catalytic characterization are described in chapter 5.2.2.

Thermal analysis

Thermal analysis was performed using a thermobalance (SSC 5200 Seiko Instruments). Gas mixture consisted of 20% O_2 and 80% He with a total gas flow in the sample chamber adjusted to 50 ml/min. Total gas flow had to be relatively low to avoid an undesired mass transport of light SBA-15 particles. Crucible for sample and reference consisted of aluminum. Temperature was kept at 303 K for 60 min and afterwards temperature was increased up to 673 K and held for 30 min. Heating rate was 10 K/min.

4.3 Structural characterization

N_2 physisorption measurements of all synthesized SBA-15 samples revealed comparable high specific surface areas, $a_{s,BET}$, and comparable pore structures. N_2 adsorption/desorption isotherms of all SBA-15 samples were of type IV with H1 hysteresis loops, indicative of mesoporous materials with regularly shaped mesopores. [68] Figure 4.1 depicts N_2

adsorption/desorption isotherms (left) and pore radius distributions (right) of three SBA-15 samples to clarify the high degree of similarity of pore structure and specific surface area of as-prepared SBA-15 samples.

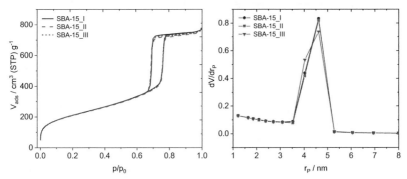

Figure 4.1: N_2 adsorption/desorption isotherms of three SBA-15 samples (left) and pore radius distribution determined by BJH method (right).

Pore radius distributions, calculated by BJH method [70], were narrow with a maximum at 4.6 nm for all SBA-15 samples. Specific surface areas, calculated by BET method [69], were high with values of 673.6–779.4 m^2 g^{-1} as expected for mesoporous SBA-15. [33, 100] Furthermore, all SBA-15 samples possessed comparable pore volumes, V_p. Pore volume ranged from 0.962 through 1.218 cm^{3}g^{-1}.

Structural properties of the SBA-15 samples were further investigated by X-ray diffraction. Wide-angle X-ray diffraction patterns of three SBA-15 samples are exemplary depicted in Figure 4.2, left. Only a broad and weak diffraction peak at 23° 2θ, corresponding to the amorphous structure of SBA-15, was observed for all samples. [49] Long-range ordered species were excluded due to the absence of further diffraction peaks. Small-angle X-ray diffraction patterns of the SBA-15 samples possessed three characteristic diffraction peaks, (10*l*), (11*l*), and (20*l*), indicative of the two-dimensional hexagonal symmetry of SBA-15 (Figure 4.2, right). [32, 33, 101] Lattice constant, a_0, of hexagonal unit cell was calculated according to equation (3.10) using peaks (10*l*), (11*l*), and (20*l*). All SBA-15 samples possessed nearly identical lattice constants ranging from 10.91 through 11.08 nm. Furthermore, wall thickness between the mesopores of SBA-15 was determined (eq. (3.11)). Values of wall thickness, d_w, were similar for all SBA-15 samples and ranged between 1.69 and 1.89 nm. Lattice constant and wall thickness for all SBA-15 samples complied with those expected for two-dimensional hexagonal ordered pore structure of SBA-15.

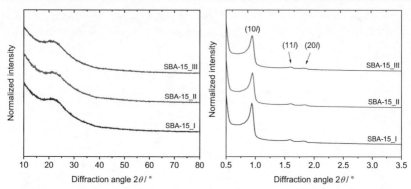

Figure 4.2: Wide-angle (left) and small-angle (right) X-ray diffraction patterns of three SBA-15 samples.

Transmission electron microscopy (TEM) micrographs of SBA-15 samples additionally confirmed two-dimensional hexagonal ordered pore structure of support material (Figure 4.3). Well-ordered hexagonal arrangement of uniform mesopores was clearly discernible. Accordingly, results from TEM measurements corroborated results from N_2 physisorption and XRD measurements. All synthesized SBA-15 samples possessed comparable structural properties indicative of a reproducible synthesis of support material. Furthermore, requirements of high specific surface area, narrow pore size distribution, and uniform pore structure for suitable support materials were fulfilled.

Figure 4.3: TEM micrographs of SBA-15. Left: Image in the direction perpendicular to the pore axis. Right: Image in the direction parallel to the pore axis.

4.4 Catalytic performance

Catalytic performance of SBA-15 support material was tested in selective oxidation of propene at 653 K in 5% propene and 5% oxygen in helium. Propene conversion during 12 h on stream was less than 0.3% as shown in Figure 4.4. Additionally, formation of selective oxidation products was not revealed. Therefore, SBA-15 support material was considered to be inactive in catalytic reaction and the requirement of low catalytic activity for suitable support materials was fulfilled.

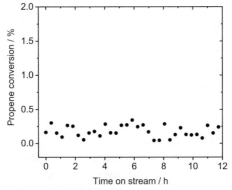

Figure 4.4: Propene conversion of SBA-15 as function of time on stream during catalytic reaction at 653 K in 5% propene and 5% oxygen in helium.

4.5 Thermal stability

Dehydration and dehydroxylation processes of SBA-15 support material were investigated by thermal analysis. Figure 4.5 depicts the thermogram of SBA-15 support material measured in 20% oxygen in helium with a heating rate of 10 K/min. Temperature was increased from room temperature up to 673 K.

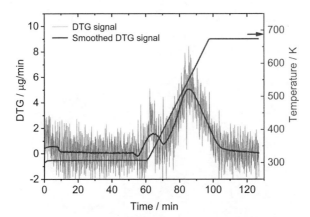

Figure 4.5: Thermogram of SBA-15 during thermal treatment in 20% oxygen in helium. Temperature was increased with 10 K/min to 673 K.

In all thermograms of SBA-15, two peaks in differential thermogravimetric curves (DTG) assigned to two different mass losses were distinguished. The first mass loss effecting a DTG peak from 307 through 398 K was attributed to the removal of physically adsorbed water on the surface of SBA-15. Temperature range of this first mass loss was in accordance with that reported by Zhuravlev [102] for dehydration process of amorphous silica. The second and more prominent peak in the thermogram of SBA-15 was observed from 403 through 673 K. Again, temperature range of the observed mass loss agreed well with that from literature. [102] Hence, the second mass loss was attributed to a dehydroxylation process where silanol groups are removed from silica surface. Total mass loss of SBA-15 during dehydration and dehydroxylation process amounted to 3.1%. Any further mass losses, possibly ascribed to decomposition of SBA-15 structure, were not observed. This indicated high thermal stability of SBA-15 support material.

4.6 Stability under reaction conditions

Investigations of structural properties of SBA-15 before and after catalytic reaction in 5% propene and 5% oxygen in helium at 653 K were conducted to ensure stability of support material under reaction conditions. Thermal stability of SBA-15 was additionally corroborated by results from comparison of structural characterization before and after catalytic reaction. Figure 4.6, left depicts N_2 adsorption/desorption isotherms of SBA-15 support material before and after catalytic reaction. Overall shape of the N_2 adsorption/desorption isotherms before and after catalytic reaction remained similar. After catalytic reaction, N_2 adsorption/desorption isotherms were still identified as type IV with H1 hysteresis loop indicative of mesoporous SBA-15. [68] Pore radius distribution (Figure 4.6, right) before and after catalytic reaction was narrow with a maximum at 4.6 nm. Hence, pore structure and pore ordering of SBA-15 remained unaffected by catalytic reaction. However, specific surface area, $a_{s,BET}$, slightly decreased after catalytic reaction accompanied by a slightly increased pore volume, V_p (Table 4.1). The decrease in specific surface area after catalytic reaction amounted to 0.8% indicative of no major structural change of SBA-15.

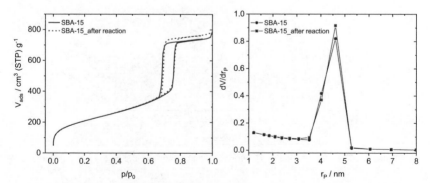

Figure 4.6: N_2 adsorption/desorption isotherms (left) and pore radius distribution (right) of SBA-15 support material before (grey) and after (blue) catalytic reaction (5% propene and 5% oxygen in helium at 653 K, 12 h time on stream).

Wide-angle (Figure 4.7, left) and small-angle (Figure 4.7, right) X-ray diffraction patterns before and after catalytic reaction confirmed that two-dimensional hexagonal ordering of SBA-15 structure was preserved. Apart from the broad and weak diffraction peak at 23° 2θ, assigned to the amorphous silica structure of SBA-15, no additional diffraction peaks were revealed. Additionally, small-angle X-ray diffraction patterns before and after catalytic reaction possessed the three characteristic diffraction peaks corresponding to SBA-15. Peak shapes and positions remained unaffected by treatment under catalytic reaction conditions. This complied with constant lattice constants and wall thickness as summarized in Table 4.1.

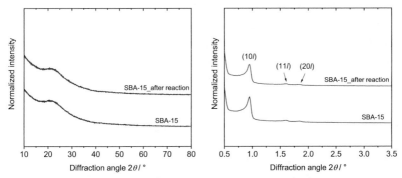

Figure 4.7: Wide-angle (left) and small-angle (right) X-ray diffraction patterns of SBA-15 support material before (grey) and after (blue) catalytic reaction (5% propene and 5% oxygen in helium at 653 K, 12 h time on stream).

Table 4.1: Structural properties of SBA-15 support material before and after catalytic reaction. Specific surface area, $a_{s,BET}$, pore volume, V_{pore}, average pore radius, r_p, lattice constant, a_0, corresponding to the hexagonal unit cell, and wall thickness, d_w, between the mesopores of SBA-15. Catalytic reaction was conducted in 5% propene and 5% oxygen in helium at 653 K.

	$a_{s,BET}$ / m^2g^{-1}	V_{pore} / $cm^3\,g^{-1}$	r_p / nm	a_0 / nm	d_w / nm
SBA-15	763.1 ± 0.6	1.176 ± 0.001	4.6	10.91 ± 0.02	1.69 ± 0.02
SBA-15_after reaction	756.8± 0.6	1.203 ± 0.001	4.6	10.91 ± 0.02	1.69 ± 0.02

4.7 Conclusion

Various SBA-15 samples were successfully synthesized and characterized by N_2 physisorption, X-ray diffraction, transmission electron microscopy, thermal analysis, and *in situ* gas chromatography combined with mass spectrometry. Characterization of SBA-15 was performed with respect to applicability as suitable support material for Fe_xO_y-based selective oxidation catalysts. All as-prepared SBA-15 samples possessed high specific surface areas and narrow pore radius distributions. Furthermore, two-dimensional hexagonal ordered pore structure with no long-range ordered species was revealed for all SBA-15 samples. Values of lattice constant of hexagonal unit cell and wall thickness between the mesopores were comparable and complied with those expected for mesoporous SBA-15. Accordingly, synthesis of SBA-15 was reproducible and synthesized SBA-15 samples showed no major differences in structural properties.

Catalytic performance of as-prepared SBA-15 was tested in selective oxidation of propene at 653 K. Propene conversion of SBA-15 during 12 h time on stream was less than 0.3% and formation of selective oxidation products was not observed. Therefore, SBA-15 was considered to be inactive in selective oxidation of propene.

Furthermore, thermal stability of SBA-15 was confirmed by thermal analysis. Thermal treatment of SBA-15 led to removal of physically adsorbed water on the surface and subsequent removal of silanol groups. Additional mass losses induced by possible decomposition reactions of SBA-15 were not revealed indicating a high thermal stability of SBA-15.

Besides thermal stability, stability of SBA-15 under reaction conditions was investigated. Pore structure of SBA-15 remained unaffected by treatment under catalytic reaction conditions. Neither lattice constant nor wall thickness varied after catalytic reaction. The slightly decreased specific surface area by 0.8% together with the slightly increased pore volume after catalytic reaction was indicative of no major structural change under reaction conditions.

Consequently, SBA-15 was proven to be a suitable support material for selective oxidation catalysts. Requirements of high specific surface area, uniform pore structure, stability, and inactivity in catalytic reaction for suitable support materials were all fulfilled. In the following, SBA-15 was used as support material for Fe_xO_y-based selective oxidation catalysts.

5 Synthesis and characterization of Fe_xO_y/SBA-15 catalysts using various precursors

5.1 Introduction

The structure of supported iron oxidic species is crucially influenced by synthesis procedure. Despite various attempts, there is still no consensus concerning the active iron oxide species in selective oxidation reactions. [48–50, 65] Dispersing iron oxides on high surface support material SBA-15 yields Fe_xO_y/SBA-15 as suitable model catalyst for investigating structure-activity correlations. For this purpose, a detailed knowledge of the catalyst structure is indispensable. Therefore, synthesized Fe_xO_y/SBA-15 were characterized with a multitude of analyzing methods, such as DR-UV-Vis, X-ray absorption, Raman and Mössbauer spectroscopy, as well as X-ray diffraction. Varying iron precursors during synthesis of Fe_xO_y/SBA-15 constituted a fundamental starting point for investigating the influence of the precursor on resulting iron oxidic species. In addition to detailed *ex situ* characterization of Fe_xO_y/SBA-15, *in situ* characterization during reduction in hydrogen and under propene oxidation conditions was conducted.

In this chapter, it will be shown that the precursor has a crucial impact on resulting iron oxidic species on SBA-15, their local structure, reducibility, and catalytic performance. Applicability of solid-state kinetic analysis to supported iron oxides will be shown. Moreover, it will be clarified that additional solid-state kinetic analysis can be helpful in corroborating structure-activity correlations of Fe_xO_y/SBA-15 catalysts. Correlation between varying precursors and local structure, reducibility, solid-state kinetic properties, and catalytic performance of Fe_xO_y/SBA-15 will be elucidated.

5.2 Experimental

5.2.1 Sample preparation

Iron oxidic species supported on SBA-15 were prepared by incipient wetness technique. The iron loading ranged between 2.0 and 10.7wt%. Therefore, an aqueous solution of $(NH_4, Fe(III))$ citrate ($\sim$ 18% Fe, Roth) or Fe(III) nitrate nonahydrate (99+%, Acros Organics) was used. After drying in air for 24 h, calcination was carried out at 723 K for 2 h. According to the iron loading and the used precursor, samples were denoted 2.5wt% Fe_Citrate, 3.0wt% Fe_Citrate, 5.5wt% Fe_Citrate, 6.3wt% Fe_Citrate, 10.7wt% Fe_Citrate, 2.0wt% Fe_Nitrate, 3.3wt% Fe_Nitrate, 5.8wt% Fe_Nitrate, 7.2wt% Fe_Nitrate, and 9.3wt% Fe_Nitrate. Furthermore, a mechanical mixture of SBA-15 and crystalline α-Fe_2O_3 (99.98%, Roth) (10.5wt% Fe) was prepared and denoted Fe_2O_3/SBA-15 (Figure 5.1). After synthesis of Fe_xO_y/SBA-15 samples, iron loadings were confirmed by X-ray fluorescence spectroscopy. Furthermore, CHN analysis was performed to confirm a complete decomposition of the used precursor. In order to clarify the influence of the precursor during synthesis of Fe_xO_y/SBA-15 samples, Fe_xO_y/SBA-15 obtained from citrate precursor were denoted citrate samples, while those obtained from nitrate precursor were denoted nitrate samples.

Uncalcined Fe_xO_y/SBA-15 samples synthesized by citrate precursor were light yellow, independent of the iron loading. After calcination the light yellow color turned to light orange brownish. For the lowest loaded sample with 2.5wt% Fe, the color was unchanged after calcination. Conversely, for the higher loaded citrate samples, the color intensity increased with increasing iron loading. Using the nitrate precursor resulted in white uncalcined Fe_xO_y/SBA-15 samples. After calcination a color change to orange for the lowest loaded sample, to reddish orange for loadings of 3.3-7.2wt% Fe, and to dark red for 9.3wt% Fe was observed. Figure 5.2 illustrates the change in color of Fe_xO_y/SBA-15 samples before and after calcination dependent on precursor and iron loading.

Figure 5.1: Mechanical mixture Fe_2O_3/SBA-15.

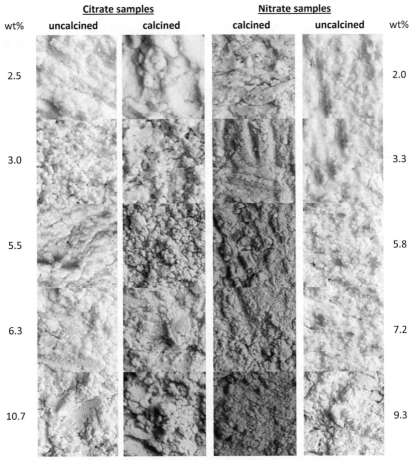

Figure 5.2: Fe_xO_y/SBA-15 samples before (uncalcined) and after (calcined) calcination dependent on used precursor. Increasing iron loading (wt%) from top to bottom.

5.2.2 Sample characterization

X-ray fluorescence analysis

Quantitative analysis of the metal oxide loadings on SBA-15 was conducted by X-ray fluorescence spectroscopy on an AXIOS X-ray spectrometer (2.4 kW model, PANalytical), equipped with a Rh K_α X-ray source, a gas flow detector, and a scintillation detector. Prior to measurements, samples were mixed with wax (Hoechst wax C micropowder, Merck), ratio 1:1, and pressed into pellets of 13 mm diameter. Quantification was performed by standardless analysis using the software package SuperQ5 (PANalytical).

CHN elemental analysis

Elemental contents of C, H, and N were determined using a Thermo FlashEA 1112 Organic Elemental Analyzer (ThermoFisher Scientific) with CHNS-O configuration. Measurements were conducted in collaboration with the measuring center at institute of chemistry at TU Berlin.

Nitrogen physisorption

Nitrogen adsorption/desorption isotherms were measured at 77 K using a BELSORP Mini II (BEL Inc. Japan). Prior to measurements, the samples were pre-treated under reduced pressure (10^{-2} kPa) at 368 K for 35 min and kept under the same pressure at 448 K for 15 h (BELPREP II vac).

Transmission electron microscopy

Transmission electron microscopy (TEM) images were recorded on a FEI Tecnai G^2 20 S-TWIN microscope equipped with a LaB_6 cathode and a 1 k x 1 k CCD camera (GATAN MS794). Acceleration voltage was set to 220 kV and samples were prepared on 300 mesh Cu grids with Holey carbon film. Measurements were conducted in collaboration with ZELMI (Zentraleinrichtung für Elektronenmikroskopie) at TU Berlin.

Powder X-ray diffraction

Powder X-ray diffraction patterns were obtained using an X'Pert PRO diffractometer (PANalytical, 40 kV, 40 mA) in theta/theta geometry equipped with a solid-state multi-channel detector (PIXel). Cu K_α radiation was used. Wide-angle diffraction scans were conducted in reflection mode. Small-angle diffraction patterns were measured in transmission mode from 0.4° through 6° 2θ in steps of 0.013° 2θ with a sampling time of 90 s/step.

Diffuse reflectance UV-Vis spectroscopy

Diffuse reflectance UV-Vis (DR-UV-Vis) spectroscopy was conducted on a two-beam spectrometer (V-670, Jasco) using a barium sulfate coated integration sphere. (Scan speed 100 nm/min, slit width 5.0 nm (UV-Vis) and 20.0 nm (NIR), and spectral region 220-2000 nm). SBA-15 was used as white standard for all samples.

Raman spectroscopy

Raman spectroscopy was performed on a Renishaw inVia Reflex Spectrometer System using a frequency doubled Nd:YAG laser (532 nm, 100 mW) and a 50x objective for focusing. Data were collected in spectral range between 60 and 1320 cm^{-1} with a resolution of 1 cm^{-1} by means of a 2400 lines mm^{-1} grating and from 79 to 4683 cm^{-1} with a resolution of 4-5 cm^{-1} by means of a 600 lines mm^{-1} grating, respectively. For each sample, five points were measured with three individual acquisitions per point, which were then averaged for improvement of the signal-to-noise ratio. The laser intensity was set to 0.5% and the exposure time was varied between 1200 and 2400 s for each acquisition, depending on the iron loading. Reference samples were measured at 0.05-0.5% laser intensity and 180-600 s exposure time. Data treatment was performed with WiRE 4.4 from Renishaw. All spectra presented were not baseline-corrected. Measurements were conducted in collaboration with the group of Jan-Dierk Grunwaldt at Karlsruhe Institute of Technology (KIT).

Mössbauer spectroscopy

Zero-field ^{57}Fe Mössbauer spectroscopic measurements were conducted on a transmission spectrometer with sinusoidal velocity sweep. Velocity calibration was done with an α-Fe foil at ambient temperature. Measurements of samples 7.2wt% Fe_Nitrate and 2.0wt% Fe_Nitrate were performed using a Janis closed-cycle cryostat with the sample container entirely immersed in helium exchange gas at 14 and 300 K. Combined with measurements over a time period of about one to twelve days the latter ensured a gradient-free sample temperature. The sample temperature was recorded with a calibrated Si diode located close to the sample container made of Teflon or PEEK (polyether ether ketone) providing a temperature stability of better than 0.1 K. Additional measurements of samples 9.3wt% Fe_Nitrate, 7.2wt% Fe_Nitrate, 2.0wt% Fe_Nitrate and 10.7wt% Fe_Citrate were carried out on a spectrometer equipped with a CryoVac continuous flow cryostat with comparable specifications, geometry, and sample environment as described above. The nominal activity of the Mössbauer sources used was about 50 mCi of ^{57}Co in a rhodium matrix. Spectra at 4 K were recorded every 30 minutes during overall measurement duration. Each Mössbauer spectrum shown here corresponds to the last spectrum in the respective series. Quantitative analysis of the recorded spectra was conducted on basis of the stochastic relaxation model developed by Blume and Tjon [103], in which the magnetic hyperfine field B_{hf} fluctuates randomly between two directions ($+B_{hf}$ and $-B_{hf}$) along the symmetry axis of an axially symmetric electric field gradient tensor. Using this model is motivated by the observation of a significant line broadening, in particular in the spectra obtained for 7.2wt% Fe_Nitrate at intermediate temperatures of approximately 60 and 100 K suggesting the presence of slow relaxation processes with relaxation times, τ_c, that are long or of the same order of magnitude as the Larmor precession time of the ^{57}Fe nuclear magnetic moment (i.e., 10^{-6} s $< \tau_c < 10^{-8}$ s). The quadrupole shift, ε, is given by $e^2 q Q/4$, assuming $e^2 q Q << \mu B_{hf}$ (constants μ, e, q, Q were used in their usual meaning). The isomer shift, δ, is reported with respect to iron metal at ambient temperature and was not corrected in terms of the second order Doppler shift. Measurements were conducted in collaboration with the group of Martin Bröring at TU Braunschweig.

X-ray absorption spectroscopy

Transmission X-ray absorption spectroscopy (XAS) was performed at the Fe K edge (7112 eV) at beamline P65 at Petra III at DESY Hamburg using a Si(111) double crystal monochromator. X-ray absorption fine structure (XAFS) scans were measured in the energy range between 7012 and 7912 eV with a scan speed of 180 s/scan. For *ex situ* measurements, references and samples were diluted with wax (Hoechst wax C micropowder, Merck) and pressed into pellets with a diameter of 13 mm. Pellets consisting of a mixture of sample and boron nitride (99.5%, Alfa Aesar) with a diameter of 5 mm were pressed for *in situ* XAS measurements. Sample masses were calculated to result in an edge jump around $\Delta\mu(d) = 1.0$ at the Fe K edge.

In situ experiments were conducted in a flow reactor at atmospheric pressure with a total flow of 40 ml/min of reactant gas mixture (5% propene, 5% oxygen in helium) in a temperature range from 298 to 653 K at a heating rate of 5 K/min. Reactant gas flow was adjusted by separate mass flow controllers (Bronkhorst) for each gas. Gas atmosphere at the outlet of the reactor was continuously analyzed using a non-calibrated mass spectrometer (Omnistar, Pfeiffer) in a multiple ion detection mode.

XAFS analysis was performed using the software package WinXAS v.3.2. [104] Background subtraction and normalization were performed by fitting a linear polynomial and a third degree polynomial to the pre-edge and the post-edge region of the X-ray absorption spectra, respectively. The extended absorption fine structure (EXAFS), $\chi(k)$, at the Fe K edge was extracted by using cubic splines to obtain a smooth atomic background, $\mu_0(k)$. The pseudo radial distribution function, $FT(\chi(k))$, was obtained by Fourier transforming the k^3-weighted experimental $\chi(k)$, multiplied by a Bessel window, into the R space. EXAFS data analysis was performed using theoretical backscattering phases and amplitudes obtained from FEFF calculations. [105] EXAFS refinements were conducted in R space to magnitude and imaginary part of the Fourier transformed k^3-weighted experimental $\chi(k)$ using the standard EXAFS formula (eq. (3.7)). EXAFS refinement procedure is described in chapter 5.3.4.

For the XANES analysis, background subtraction and normalization were performed by fitting linear polynomials to the pre-edge and post-edge region of the X-ray absorption spectra in the energy range from 7.06 through 7.34 keV. Further details on XANES analysis are described in chapter 5.3.4.

Temperature-programmed reduction

Temperature-programmed reduction (TPR) was performed using a BELCAT_B (BEL Inc. Japan). Samples were placed on silica wool in a silica glass tube reactor. Evolving water was trapped using a molecular sieve (4 Å). Gas mixture consisted of 5% H_2 in 95% Ar with a total gas flow of 40 ml/min. Heating rates used were 5, 10, 15, and 20 K/min to 1223 K. A constant initial sample weight of 0.03 g was used and H_2 consumption was continuously monitored by a thermal conductivity detector.

Catalytic characterization

Quantification of catalytic activity in selective oxidation of propene was performed using a laboratory fixed-bed reactor connected to an online gas chromatography system (CP-3800, Varian) and a mass spectrometer (Omnistar, Pfeiffer Vacuum). The fixed-bed reactor consisted of a SiO_2 tube (30 cm length, 9 mm inner diameter) placed vertically in a tube furnace. A P3 frit was centered in the SiO_2 tube in the isothermal zone, where the sample was placed. The samples were diluted with boron nitride (99.5%, Alfa Aesar) to achieve a constant volume in the reactor and to minimize thermal effects. Overall sample masses were 0.25 g. The reactor was operated at low propene conversion levels (5-10%) to ensure differential reaction conditions. For catalytic testing in selective oxidation of propene, a gas mixture of 5% propene and 5% oxygen in helium was used in a temperature range of 298-653 K with a heating rate of 5 K/min. Reactant gas flow rates of propene, oxygen, and helium were adjusted through separate mass flow controllers (Bronkhorst) to a total flow of 40 ml/min. All gas lines and valves were preheated to 473 K. Hydrocarbons and oxygenated reaction products were analyzed using a Carbowax 52CB capillary column, connected to an Al_2O_3/MAPD capillary column (25 m x 0.32 mm) or a fused silica restriction (25 m x 0.32 mm), each connected to a flame ionization detector (FID). Permanent gases (CO, CO_2, N_2, O_2) were separated and analyzed using a "Permanent Gas Analyzer" (CP-3800, Varian) with a Hayesep Q (2 m x 1/8") and a Hayesep T packed column (0.5 m x 1/8") as precolumns combined with a back flush. For separation, a Hayesep Q packed column (0.5 m x 1/8") was connected via a molecular sieve (1.5 m x 1/8") to a thermal conductivity detector (TCD). Additionally, product and reactant gas flow were continuously monitored by a connected mass spectrometer (Omnistar, Pfeiffer) in a multiple ion detection mode.

Based on the measured volume fractions, *Vol%*, conversion, *X*, of the key component propene, and selectivity, *S*, towards a desired product, *i*, were calculated according to:

$$X_{propene} = \frac{Vol\%_{propene,in} - Vol\%_{propene,out}}{Vol\%_{propene,in}} \tag{5.1}$$

$$S_i = \frac{a_i}{a_{propene}} \cdot \frac{Vol\%_{i,out} - Vol\%_{i,in}}{\sum_i \left(\frac{a_i}{a_{Propen}} \cdot Vol\%_{px,out} \right)} \tag{5.2}$$

with number of carbon atoms in the desired product, a_i, and in propene, $a_{propene}$ ($a_{propene}$ = 3). The sum of the volume fractions of all measured products, *px*, is considered for the calculation of the selectivity towards a desired product. Determining the carbon atom balances according to equation (5.3) was required to ensure reasonable results of the calculated selectivities.

$$\frac{Vol\%(C)_{out}}{Vol\%(C)_{in}} = \frac{\sum(a_{px} \cdot Vol\%_{px,out} + 3 \cdot Vol\%_{propene,out})}{3 \cdot Vol\%_{propene,in}} \tag{5.3}$$

For all samples, carbon atom balances were higher than 0.98. Reaction rate for a desired product, r_i, was calculated by

$$r_i = \frac{X_{propene} \cdot Vol\%_{propene,in} \cdot Vol\%_i}{m_{Kat} \cdot V_m \cdot 60 \frac{s}{min}} \tag{5.4}$$

with effective catalyst mass, m_{cat}, and molar gas volume, V_m.

5.3 Structural characterization of Fe$_x$O$_y$/SBA-15 catalysts

5.3.1 Pore structure and surface properties

Iron oxide dispersion

Pore structure and surface properties of Fe$_x$O$_y$/SBA-15 samples were determined by nitrogen physisorption and transmission electron microscopy (TEM) measurements. Nitrogen adsorption/desorption isotherms of Fe$_x$O$_y$/SBA-15 samples at various iron loadings dependent on used precursor are depicted in Figure 5.3. All Fe$_x$O$_y$/SBA-15 samples exhibited type IV N$_2$ adsorption/desorption isotherms indicating mesoporous materials. Adsorption and desorption branches were nearly parallel at the hysteresis loop, as expected for regularly shaped pores. Hysteresis loops of the Fe$_x$O$_y$/SBA-15 samples were of type H1, according to IUPAC. [68]

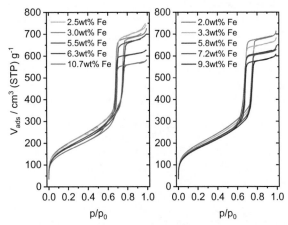

Figure 5.3: N$_2$ adsorption/desorption isotherms of Fe$_x$O$_y$/SBA-15 samples. Left: Citrate samples. Right: Nitrate samples.

Both SBA-15 and Fe$_x$O$_y$/SBA-15 samples exhibited high specific surface areas with narrow pore size distributions. Independent of the precursor, low loaded Fe$_x$O$_y$/SBA-15 samples showed significantly higher specific surface areas compared to higher loaded samples. All Fe$_x$O$_y$/SBA-

15 samples showed a decrease in specific surface area compared to corresponding SBA-15. Specific surface area, $a_{s,BET}$, of SBA-15 ranged between 673.6 and 779.4 m^2g^{-1}, whereas that of the citrate samples ranged from 605.7 through 725.2 m^2g^{-1} and that of the nitrate samples from 633.1 through 703.0 m^2g^{-1} (Table 5.1).

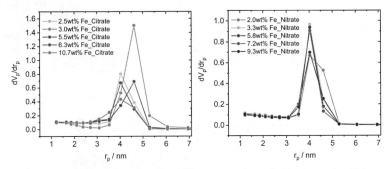

Figure 5.4: Pore radius distribution of Fe_xO_y/SBA-15 samples. Left: Citrate samples. Right: Nitrate samples.

Pore radius distribution was calculated by BJH method. [70] Fe_xO_y/SBA-15 samples showed a narrow pore radius distribution with a main maximum at 4.0 nm and 4.6 nm, respectively (Figure 5.4). Furthermore, pore volume, V_{pore}, was determined for SBA-15 and corresponding Fe_xO_y/SBA-15 samples. Supporting iron oxides on SBA-15 induced a decrease in pore volume towards 0.898-1.142 cm^3g^{-1} compared to corresponding SBA-15 with values ranging from 0.962 through 1.218 cm^3g^{-1}. The decrease in specific surface area, $a_{s,BET}$, together with the decrease in pore volume, V_{pore}, of as-prepared samples, indicated the presence of iron oxidic species in the mesopores of SBA-15. Moreover, transmission electron microscopy measurements of Fe_xO_y/SBA-15 samples also indicated that iron oxidic species were located in the pore system of SBA-15 with no iron oxidic species detected on the external surface of SBA-15 (Figure 5.5). TEM micrographs of both Fe_xO_y/SBA-15 samples and SBA-15 showed characteristic hexagonal ordering of mesopores in the two-dimensional framework of SBA-15. This was in conjunction with identified type IV N_2 adsorption/desorption isotherms with H1 hysteresis loop (Figure 5.3) for Fe_xO_y/SBA-15 samples, and hence, confirmed SBA-15 structure preservation after synthesis. In contrast to Fe_xO_y/SBA-15 samples, TEM micrograph of the mechanical mixture clearly indicated the presence of α-Fe_2O_3 crystallites beside characteristic SBA-15 pore structure (Figure 5.6). Absence of observable Fe_xO_y particles in TEM micrographs of Fe_xO_y/SBA-15 samples (Figure 5.5) indicated the iron oxidic species being highly dispersed, small, i.e. smaller than 5 nm, and in the pore system of SBA-15.

SBA-15	9.3wt% Fe_Nitrate	10.7wt% Fe_Citrate

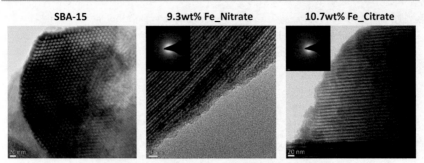

Figure 5.5: TEM micrographs of SBA-15 (left), 9.3wt% Fe_Nitrate (middle), and 10.7wt% Fe_Citrate (right). Inset depicts small-angle electron diffraction (SAED) image.

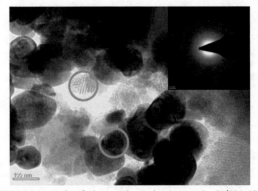

Figure 5.6: TEM micrograph of the mechanical mixture, Fe_2O_3/SBA-15. Blue circle: characteristic SBA-15 pore structure. Green circle: α-Fe_2O_3 particle. Inset depicts small-angle electron diffraction (SAED) image.

Iron oxide surface coverage, summarized in Table 5.1, varied dependent on iron loading from 0.30 through 1.90 Fe atoms per nm^2. These comparatively low values coincided with the observations reported by Wachs et al. [106]. They reported metal oxide surface coverages for silica support material being always significantly lower than theoretical monolayer capacity. Accordingly, lower surface coverages of metal oxides on silica support material resulted from lower density and reactivity of surface hydroxyl groups compared to other support materials. Investigations of surface coverage of vanadium oxide species on various support materials revealed a maximum surface coverage of approximately 2-3 V atoms per nm^2 for silica support material being below monolayer coverage. [107–109] Assuming a comparable theoretical space requirement of V_xO_y and Fe_xO_y enabled a classification of determined iron oxide surface coverage. Since iron oxide surface coverage of all Fe_xO_y/SBA-15 samples was lower than 2-3 Fe atoms per nm^2, it was considered to be below monolayer coverage, complying with the results from TEM micrographs and N_2 physisorption.

Table 5.1: Surface and porosity characteristics of Fe_xO_y/SBA-15 samples. Specific surface area, $a_{s,BET}$, pore volume, V_{pore}, ratio of mesopore surface area, a_{pore}, and specific surface area, $a_{s,BET}$, as measure of micropore contribution to the entire surface of SBA-15, differences in fractal dimension, ΔD_f, between SBA-15 and corresponding Fe_xO_y/SBA-15 samples as measure of the roughness of the surface, and surface coverage, $\Phi_{Fe\ atoms}$.

	$a_{s,BET}$ / m² g⁻¹	V_{pore} / cm³ g⁻¹	$a_{pore}/a_{s,BET}$	ΔD_f	$\Phi_{Fe\ atoms}$ / Fe atoms nm⁻²
2.5wt% Fe_Citrate	725.2 ± 0.7	1.112 ± 0.001	0.88	0.12 ± 0.02	0.37
3.0wt% Fe_Citrate	682.5 ± 0.7	1.142 ± 0.001	0.92	0.15 ± 0.02	0.47
5.5wt% Fe_Citrate	670.4 ± 0.7	1.081 ± 0.001	0.92	0.19 ± 0.05	0.88
6.3wt% Fe_Citrate	650.9 ± 0.7	0.968 ± 0.001	0.89	0.21 ± 0.05	1.04
10.7wt% Fe_Citrate	605.7 ± 0.6	0.898 ± 0.001	0.91	0.15 ± 0.01	1.90
2.0wt% Fe_Nitrate	703.0 ± 0.7	1.096 ± 0.001	0.88	0.06 ± 0.01	0.30
3.3wt% Fe_Nitrate	682.2 ± 0.7	1.049 ± 0.001	0.89	0.05 ± 0.02	0.52
5.8wt% Fe_Nitrate	654.5 ± 0.7	0.987 ± 0.001	0.87	0.04 ± 0.03	0.95
7.2wt% Fe_Nitrate	647.8 ± 0.6	0.971 ± 0.001	0.88	0.07 ± 0.01	1.19
9.3wt% Fe_Nitrate	633.1 ± 0.6	0.939 ± 0.001	0.86	-0.06 ± 0.02	1.58

Contribution of micropores to entire surface area

Mesopore surface area, a_{pore}, calculated by BJH method [70] and specific surface area, $a_{s,BET}$, calculated by BET method [69], can be used for estimating the contribution of micropores to the entire surface area of SBA-15. [71, 110] Therefore, the ratio of mesopore surface area and specific surface area was calculated (Table 5.1). A higher ratio of $a_{pore}/a_{s,BET}$ was observed for Fe_xO_y/SBA-15 samples obtained from citrate precursor compared to those obtained from nitrate precursor. The higher ratio of $a_{pore}/a_{s,BET}$ for the citrate samples indicated a lower contribution of micropores to the entire surface area of SBA-15, and thus, a higher degree of micropore filling. Apparently, using the citrate precursor resulted in stronger interactions between SBA-15 and Fe(III) atoms in the precursor, and hence, in a higher degree of micropore filling. However, lowest loaded citrate and nitrate sample possessed a ratio of $a_{pore}/a_{s,BET}$ of 0.88. This indicated an equal contribution of micropores to entire surface area for the lowest iron loadings independent of the precursor.

Surface roughness

In addition to BET and BJH method, modified FHH method was used to analyze the nitrogen physisorption data. Herein, the fractal dimension D_f was determined as a measure of the roughness of the surface. [71, 72] For all Fe_xO_y/SBA-15 samples and SBA-15, the fractal dimension ranged between 2 and 3, indicating a rough surface. To elucidate the effect of supported iron oxidic species on surface roughness of the support material, ΔD_f values were calculated as difference between D_f values of SBA-15 and those of Fe_xO_y/SBA-15 samples. In contrast to the nitrate samples, citrate samples possessed significantly higher values of ΔD_f

(Table 5.1, Figure 5.7). Therefore, compared to those of the nitrate samples, surfaces of the citrate samples appeared to be smoother. A smoother surface of the citrate samples was in accordance with the lower contribution of micropores to entire surface area for these samples. A possible explanation for the differences in surface roughness of the support material dependent on precursor might be the different strengths of chelating effect of the two Fe(III) precursors. The citrate precursor possessed a more pronounced chelating effect, and therefore, stronger bonds between citrate ligands and Fe(III) central atoms. Due to these stronger bonds, polydentate citrate ligands encapsulated the Fe(III) ions, thereby preventing agglomeration of iron oxidic species during calcination. Thus, after calcination and removal of the citrate ligands, the resulting Fe(III) species were more isolated and dispersed on the support material. This was accompanied by a more pronounced micropore filling, and therefore, a smoother surface of support material. Conversely, nitrate ligands showed minor interactions with the Fe(III) ions due to the less pronounced chelating effect. Therefore, nitrate removal during calcination was facilitated and the resulting Fe(III) species readily aggregated and formed less dispersed iron oxidic species on the support material. Consequently, compared to corresponding citrate samples, a higher amount of micropores remained unfilled yielding a higher surface roughness.

Highest loaded nitrate sample, 9.3wt% Fe_Nitrate, possessed the lowest value of ΔD_f, and hence, a rougher surface compared to all other Fe_xO_y/SBA-15 samples (Figure 5.7). Lowest ΔD_f for this sample complied with lowest ratio of $a_{pore}/a_{s,BET}$ of 0.86 (Table 5.1). For the nitrate precursor, a high iron loading resulted in less dispersed and higher aggregated iron oxidic species, being too large to fill the micropores of SBA-15. Therefore, contribution of micropores to entire surface area remained larger while surface was rougher compared to lower loaded samples.

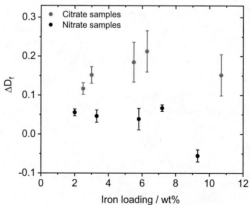

Figure 5.7: Differences in fractal dimension, ΔD_f, between SBA-15 support material and corresponding Fe_xO_y/SBA-15 samples, dependent on precursor and iron loading.

5.3.2   Long-range ordered structure of Fe_xO_y/SBA-15

Long-range ordered structure characterization of the Fe_xO_y/SBA-15 samples was performed using wide-angle and small-angle powder X-ray diffraction. Figure 5.8 depicts wide-angle diffraction patterns of the Fe_xO_y/SBA-15 samples, the mechanical mixture Fe_2O_3/SBA-15, and crystalline α-Fe_2O_3. For all Fe_xO_y/SBA-15 samples, a broad diffraction peak at 23° 2θ, corresponding to the amorphous silica structure of SBA-15 [49], was observed. The absence of further diffraction peaks indicated highly dispersed, small, and isolated iron oxidic species supported on SBA-15. Conversely, XRD pattern of the mechanical mixture showed diffraction peaks of crystalline α-Fe_2O_3.

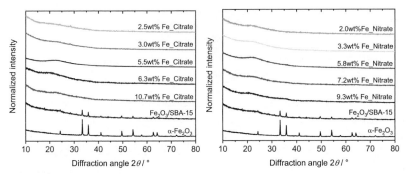

Figure 5.8: Wide-angle X-ray diffraction patterns of Fe_xO_y/SBA-15 samples (left: citrate samples, right: nitrate samples), mechanical mixture of α-Fe_2O_3 and SBA-15, and crystalline α-Fe_2O_3.

Small-angle diffraction patterns of all Fe_xO_y/SBA-15 samples exhibited characteristic diffraction peaks (10l), (11l), and (20l), corresponding to the two-dimensional hexagonal symmetry of SBA-15 (Figure 5.9). [32, 33, 101] Hence, SBA-15 structure was preserved after supporting iron oxidic species on SBA-15. However, a slight shift of the characteristic diffraction peaks towards lower values of 2θ compared to support material was observed. Lattice constant, a_0, of the hexagonal unit cell was determined from the diffraction peaks (10l), (11l), and (20l), using equation (3.10) for all Fe_xO_y/SBA-15 samples. Table 5.2 summarizes values of lattice constant and wall thickness.

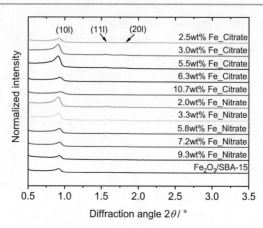

Figure 5.9: Small-angle X-ray diffraction patterns of Fe_xO_y/SBA-15 samples and mechanical mixture of α-Fe_2O_3 and SBA-15 (Fe_2O_3/SBA-15).

Independent of the precursor, supporting iron oxides on SBA-15 induced an increased lattice constant, a_0, of 10.94-11.16 nm and wall thickness, d_w, of 2.96-3.10 nm. SBA-15 possessed values of lattice constant of 10.91-11.08 nm and values of wall thickness of 1.69-1.86 nm. The increase in lattice constant and wall thickness due to supporting iron oxides on SBA-15 indicated a slight expansion of the hexagonal unit cell of SBA-15. Since Fe^{3+} ions possess a larger ion radius compared to Si^{4+} ions, partially incorporated Fe^{3+} in the mesoporous framework of SBA-15 probably effected the increase of a_0. [111] This confirmed a successful insertion of iron oxidic species in the mesopores, attached to the walls, of SBA-15. Furthermore, an influence of the precursor was revealed. The nitrate samples possessed slightly higher values of lattice constant and wall thickness compared to corresponding citrate samples with similar iron loading (Table 5.2). This might be indicative of larger iron oxidic species obtained from nitrate precursor. Precursor dependence of lattice constant and wall thickness was more distinct for iron loadings higher than 5.5wt%. However, values of lattice constant of as-prepared Fe_xO_y/SBA-15 samples were comparable to those of SBA-15, and hence, SBA-15 pore structure was retained after synthesis of Fe_xO_y/SBA-15 samples.

Table 5.2: Lattice constant, a_0, and wall thickness, d_w, of Fe_xO_y/SBA-15 samples.

	a_0 / nm	d_w / nm		a_0 / nm	d_w / nm
2.5wt% Fe_Citrate	11.15 ± 0.02	3.09 ± 0.02	2.0wt% Fe_Nitrate	11.16 ± 0.01	3.10 ± 0.01
3.0wt% Fe_Citrate	11.10 ± 0.02	3.03 ± 0.02	3.3wt% Fe_Nitrate	11.14 ± 0.02	3.09 ± 0.02
5.5wt% Fe_Citrate	11.08 ± 0.03	3.00 ± 0.03	5.8wt% Fe_Nitrate	11.12 ± 0.02	3.06 ± 0.02
6.3wt% Fe_Citrate	11.02 ± 0.02	2.96 ± 0.02	7.2wt% Fe_Nitrate	11.11 ± 0.02	3.05 ± 0.02
10.7wt% Fe_Citrate	10.94 ± 0.02	2.89 ± 0.02	9.3wt% Fe_Nitrate	11.10 ± 0.01	3.03 ± 0.01

5.3.3 Diffuse reflectance UV-Vis, Raman, and Mössbauer spectroscopy

DR-UV-Vis spectroscopy

DR-UV-Vis spectra of Fe_xO_y/SBA-15 samples dependent on precursor and iron loading are depicted in Figure 5.10. All samples showed a broad absorption resulting from overlapping absorption bands. The intense and broad absorption of all Fe_xO_y/SBA-15 samples was ascribed to charge transfer (CT) transitions between oxygen ligands (O^{2-}) and central iron atoms (Fe^{3+}) in iron oxidic species on SBA-15. Symmetry and spin forbidden single electron d-d transitions in Fe^{3+} leading to weak absorption bands at low wavenumber range (< 20000 cm^{-1}) were not observed. [112, 113] Hence, Fe_xO_y/SBA-15 samples possessed symmetric coordination spheres, independent of both iron loading and precursor.

For lowest loaded and highest loaded citrate and nitrate samples, DR-UV-Vis spectra before calcination are exemplary depicted in Figure 5.10. Comparing DR-UV-Vis spectra of uncalcined and corresponding calcined Fe_xO_y/SBA-15 samples revealed a red-shift of the absorption after calcination. Independent of the precursor, red-shift of the absorption was significantly more pronounced for higher iron loadings. Furthermore, nitrate samples showed a more pronounced red-shift of the absorption after calcination compared to corresponding citrate samples. This was indicative of a different change in coordination sphere and electronic structure of the iron oxidic species dependent on the precursor. Apparently, less pronounced chelating effect of the nitrate precursor (see chapter 5.3.1) induced a stronger red-shift of the absorption after calcination, and hence, a more pronounced change in electronic structure and coordination sphere of iron. Additionally, calcination of Fe_xO_y/SBA-15 samples induced a broadening of the absorption indicative of various coexisting coordination spheres.

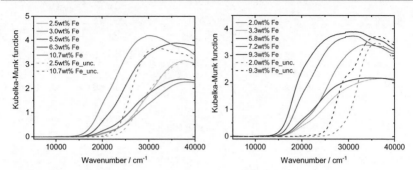

Figure 5.10: DR-UV-Vis spectra of citrate samples (left) after calcination (straight lines) and of nitrate samples (right) after calcination (straight lines). DR-UV-Vis spectra before calcination (dashed lines) are exemplary depicted for lowest loaded and highest loaded Fe$_x$O$_y$/SBA-15 samples.

In the following, results of DR-UV-Vis spectroscopy are presented for calcined Fe$_x$O$_y$/SBA-15 samples. Independent of the precursor, increasing iron loading induced a red-shift of the absorption accompanied by a broadening. This effect of increasing iron loading resulted from both increasing number of oxygen ligands and increasing aggregation of iron oxidic species on SBA-15. DR-UV-Vis edge energies, determined from DR-UV-Vis spectra according to Weber [114], were applied for estimating average iron oxide particle size of Fe$_x$O$_y$/SBA-15 samples. Plotting $[F(R_\infty)*h\nu]^2$ as function of $h\nu$ and fitting a straight line through low energy rise in the plot resulted in DR-UV-Vis edge energy at the energy intercept of the straight line. In general, a correlation between decreasing DR-UV-Vis edge energy and increasing average particle size of transition metal oxides based on the theoretical model of a particle in the box is proposed. [49, 114, 115] DR-UV-Vis edge energies of Fe$_x$O$_y$/SBA-15 samples and mechanical mixture Fe$_2$O$_3$/SBA-15 were determined and depicted as function of iron loading (Figure 5.11). Significant differences in DR-UV-Vis edge energy were observed dependent on the precursor. The citrate precursor led to significantly higher DR-UV-Vis edge energies compared to those obtained from the nitrate precursor. Hence, citrate precursor yielded significantly smaller iron oxidic species compared to nitrate precursor, indicative of a precursor-dependent particle size effect. Accordingly, citrate precursor induced a higher iron oxide dispersion, accompanied by a lower aggregation and a smaller average species size. This precursor-dependent particle size effect was in accordance with the differences in surface and porosity characteristics of Fe$_x$O$_y$/SBA-15 samples due to the different strengths of chelating effect of the two Fe(III) precursors (see chapter 5.3.1).

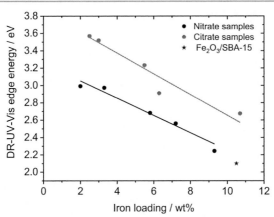

Figure 5.11: DR-UV-Vis edge energy of Fe_xO_y/SBA-15 samples and mechanical mixture Fe_2O_3/SBA-15 as function of iron loading.

Independent of the precursor, DR-UV-Vis edge energy of Fe_xO_y/SBA-15 samples decreased with increasing iron loading, indicative of an increasing average iron oxide species size. Hence, increasing iron loading led to an increased aggregation and less dispersion of iron oxidic species on SBA-15. However, edge energy values of Fe_xO_y/SBA-15 samples were higher than that of crystalline α-Fe_2O_3 (2.10 eV) as observed for the mechanical mixture. This showed that iron oxidic species on SBA-15 were smaller than crystallites of α-Fe_2O_3 and agreed with the results from XRD measurements (see chapter 5.3.2). Edge energy values of the citrate samples ranged from 3.56 eV for 2.5wt% Fe_Citrate through 2.67 eV for 10.7wt% Fe_Citrate, whereas those of the nitrate samples ranged from 2.99 eV for 2.0wt% Fe_Nitrate through 2.24 eV for 9.3wt% Fe_Nitrate. To further elucidate the coordination environment of Fe_xO_y/SBA-15 samples, DR-UV-Vis edge energy values were compared to those of reference compounds. Fe(III) nitrate precursor, consisting of isolated octahedrally coordinated Fe^{3+}, possessed an edge energy of 3.53 eV, whereas edge energy of isolated tetrahedrally coordinated Fe^{3+} in Fe-MFI zeolite (Si/Fe = 64) is reported to be 4.68 eV. [116] DR-UV-Vis edge energies of Fe_xO_y/SBA-15 samples were lower than that expected for isolated tetrahedrally coordinated iron species, and higher than that of polymeric octahedrally coordinated Fe^{3+}, as present in crystalline α-Fe_2O_3. All Fe_xO_y/SBA-15 samples, except 2.5wt% Fe_Citrate, possessed higher edge energy values than 3.53 eV. Hence, octahedral coordination environment of iron oxidic species on SBA-15 seemed reasonable.

In order to differentiate between overlapping absorption bands of the Fe_xO_y/SBA-15 samples, DR-UV-Vis spectra were fitted with Gaussian functions. Considered spectral region ranged from 5000 through 40000 cm^{-1}. Full width at half maximum (FWHM) of all fitted Gaussian functions were correlated to be equal. Furthermore, F-Tests were performed for each individual fitting procedure for confirming statistical certainty of fitting parameters. Fitting DR-UV-Vis spectra of highest loaded nitrate sample, 9.3wt% Fe_Nitrate, and mechanical mixture required four Gaussian functions, whereas for all other Fe_xO_y/SBA-15 samples, only three Gaussian functions were sufficient to describe the broad absorption. Positions of fitted Gaussian functions were used to assign corresponding absorption band to an electronic transition.

For iron oxidic species, characteristic absorption bands can be classified. Intense absorption bands in high wavenumber range (> 20000 cm^{-1}) are commonly attributed to charge transfer transitions. Absorption bands at wavenumbers above 41000 cm^{-1} are attributed to CT transitions of isolated tetrahedrally coordinated Fe^{3+}. [48–50, 112] CT transitions of O^{2-} ligands to isolated octahedrally coordinated Fe^{3+} lead to absorption bands at wavenumbers higher than 33333 cm^{-1}. [48, 112, 117] Furthermore, absorption bands resulting from CT transitions of octahedrally coordinated Fe^{3+} in dimeric or small oligomeric Fe_xO_y are generally expected at wavenumbers between 25000 and 33333 cm^{-1}. Position of these CT transitions is dependent on coordination number of oxygen ligands and degree of aggregation. Increasing number of oxygen ligands is reported to yield a red-shift of corresponding absorption band. [48, 112, 113] Moreover, Arena et al. [50] postulated a correlation between decreasing electronic absorption energy of CT absorption band of octahedrally coordinated Fe^{3+} in Fe_xO_y oligomers and increasing oligomer size. Absorption bands at wavenumbers below 25000 cm^{-1} are mainly attributed to CT transitions in crystalline α-Fe_2O_3 particles. [48, 50, 112] However, other iron oxides, e.g. FeOOH, also show absorption bands in this wavenumber region. In addition to intense absorption bands resulting from charge transfer transitions, weak absorptions bands of symmetry and spin forbidden d-d transitions might be observed at low wavenumber region below 20000 cm^{-1}. Despite these transitions being symmetry and spin forbidden, they may occur with a definite transition possibility. Furthermore, these transitions are assisted by magnetic coupling of electronic spins of two neighboring Fe^{3+} ions. Distinction between different iron oxides can sometimes be done based on absorption bands of d-d transitions, i.e. d-d electron pair transitions. For α-Fe_2O_3, d-d electron pair transitions are expected at wavenumbers up to 19230 cm^{-1}, while d-d electron pair transitions of edge-shared octahedra, as in FeOOH, are expected at wavenumbers higher than 19230 cm^{-1}. [112, 113, 118, 119]

Figure 5.12 depicts DR-UV-Vis spectra of Fe_xO_y/SBA-15 samples together with fitted Gaussian functions and corresponding sum functions. None of the Fe_xO_y/SBA-15 samples showed absorption bands attributed to d-d transitions. Hence, supporting iron oxides on SBA-15 resulted in symmetric coordination environment of Fe^{3+} ions. Absorption bands at wavenumbers lower than 25000 cm^{-1}, resulting from CT transitions in α-Fe_2O_3 particles, were only observed for highest loaded citrate sample, 10.7wt% Fe_Citrate, highest loaded nitrate sample, 9.3wt% Fe_Nitrate, and 7.2wt% Fe_Nitrate. For all lower loaded citrate and nitrate samples, first Gaussian function was ascribed to CT transitions of octahedrally coordinated Fe^{3+} in small Fe_xO_y oligomers. This was indicative of a particle size effect dependent on iron loading. Higher iron loadings resulted in small fractions of α-Fe_2O_3 particles, being however significantly smaller than crystalline α-Fe_2O_3, as confirmed by XRD results (see chapter 5.3.2).

Conversely, lower iron loadings induced small Fe_xO_y oligomers. Second Gaussian function of all Fe_xO_y/SBA-15 samples was attributed to CT transitions in small Fe_xO_y oligomers. Fe_xO_y oligomers of lower loaded Fe_xO_y/SBA-15 samples ascribed to this second absorption band were smaller than those considered by the first absorption band. Third Gaussian function in DR-UV-Vis spectra of all Fe_xO_y/SBA-15 samples was located at wavenumbers higher than 33333 cm^{-1} but lower than 41000 cm^{-1}. Hence, this Gaussian function was assigned to CT transitions of isolated octahedrally coordinated Fe^{3+} species. In contrast to all other Fe_xO_y/SBA-15 samples, highest loaded nitrate sample required four Gaussian functions for satisfactory refinement of the broad absorption. These four Gaussian functions were attributed to the presence of small α-Fe_2O_3 particles, Fe_xO_y oligomers with two different sizes, and isolated octahedrally coordinated Fe^{3+} species supported on SBA-15.

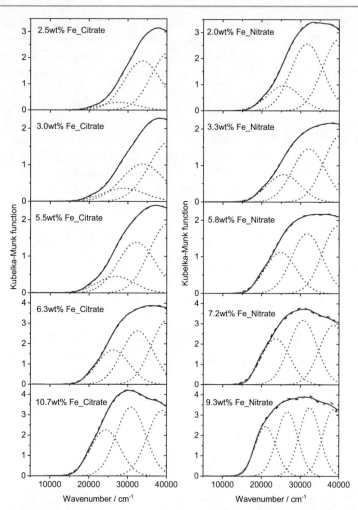

Figure 5.12: Refinement of sum function (dashed lines) of fitted Gaussian functions (dotted lines) to experimental DR-UV-Vis spectra (straight lines) of Fe_xO_y/SBA-15 samples with increasing iron loading from top to bottom. Left: Citrate samples. Right: Nitrate samples.

Positions of Gaussian functions in experimental DR-UV-Vis spectra of Fe_xO_y/SBA-15 samples as function of iron loading showed distinct differences dependent on both iron loading and precursor (Figure 5.13). Absorption band of CT transitions in Fe_xO_y oligomers, expressed by Gaussian function in wavenumber range from 25000 through 33333 cm^{-1}, was red-shifted with increasing iron loading (Figure 5.13). Hence, highest loaded Fe_xO_y/SBA-15 samples possessed largest Fe_xO_y oligomers being in accordance with revealed particle size effect. An increasing size of Fe_xO_y oligomers or small α-Fe_2O_3 particles with increasing iron loading was furthermore revealed due to the red-shift of the absorption band at wavenumbers lower than 30000 cm^{-1}. 10.7wt% Fe_Citrate showed an absorption band at 24305 cm^{-1}, whereas those of 7.2wt% Fe_Nitrate and 9.3wt% Fe_Nitrate were located at 23625 and 21061 cm^{-1}, respectively. Hence, size of small α-Fe_2O_3 particles increased from 10.7wt% Fe_Citrate to 7.2wt% Fe_Nitrate, and further for 9.3wt% Fe_Nitrate. Absorption band corresponding to CT transitions of isolated octahedrally coordinated Fe^{3+} was slightly blue-shifted from lowest loaded citrate sample to 3.0wt% Fe_Citrate and from lowest loaded nitrate sample to 3.3wt% Fe_Nitrate. For iron loadings higher than 3.3wt%, this absorption band was red-shifted. Zhang et al. [48] reported a correlation between increasing number of oxygen ligands and red-shifted absorption band of CT transition in isolated octahedrally coordinated Fe^{3+}. Based on this correlation, increasing iron loading up from ~ 4wt% resulted in an increased number of oxygen ligands around Fe^{3+}.

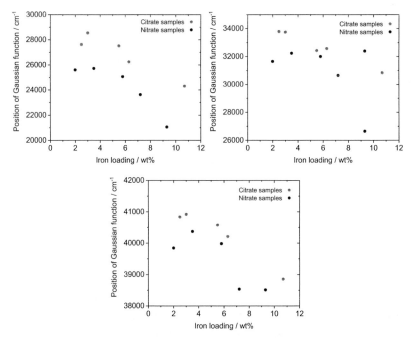

Figure 5.13: Position of Gaussian functions in experimental DR-UV-Vis spectra of Fe_xO_y/SBA-15 samples as function of iron loading.

Additionally, used precursor had a significant influence on absorption bands in DR-UV-Vis spectra of Fe_xO_y/SBA-15 samples. Using the citrate precursor resulted in a blue-shift of all considered absorption bands compared to corresponding nitrate samples (Figure 5.13). This blue-shift was indicative of a precursor-dependent particle size effect. Hence, smaller iron oxidic species were obtained from the citrate precursor compared to the nitrate precursor.

This was in accordance with surface and porosity characteristics of Fe_xO_y/SBA-15 samples (see chapter 5.3.1).

DR-UV-Vis spectra of the mechanical mixture, Fe_2O_3/SBA-15, was furthermore fitted with four Gaussian functions (Figure 5.14). However, absorption bands were located in a significantly lower wavenumber range compared to Fe_xO_y/SBA-15 samples, as expected for α-Fe_2O_3 particles. First and second Gaussian function at 11518 and 15208 cm^{-1} were ascribed to d-d transitions resulting from d orbital splitting in Fe d^5 electronic configuration. [118, 119] Third Gaussian function being located at 18469 cm^{-1} indicated ligand field transitions of octahedrally coordinated Fe^{3+}. These three absorption bands at wavenumbers lower than 19230 cm^{-1} were characteristic for α-Fe_2O_3 particles. [112, 119] Fourth Gaussian function at 21486 cm^{-1}, assigned to CT transitions in α-Fe_2O_3 particles, furthermore, indicated the presence of crystalline α-Fe_2O_3 particles being mixed with SBA-15 in the mechanical mixture.

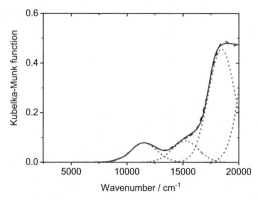

Figure 5.14: Refinement of sum function (dashed lines) of fitted Gaussian functions (dotted lines) to experimental DR-UV-Vis spectra (straight lines) of Fe_2O_3/SBA-15.

Raman spectroscopy

Raman spectroscopy was performed as complementary analyzing method to DR-UV-Vis spectroscopy for determining the influence of the precursor on type of resulting iron oxidic species. To ensure a comparability of iron oxidic species with various iron oxide references, Raman spectra were recorded for bulk α-Fe_2O_3, γ-Fe_2O_3, Fe_3O_4, and FeO (Figure A.12.1). Assigned Raman bands of recorded reference spectra agreed with those presented by Faria et al. [120] and Nasibulin et al. [121]. Furthermore, additional reference spectra of β-FeOOH were recorded. Figure 5.15 depicts Raman spectra of Fe_xO_y/SBA-15 samples at lowest, middle, and highest iron loadings, together with that of the mechanical mixture Fe_2O_3/SBA-15.

All Fe_xO_y/SBA-15 samples and the mechanical mixture exhibited Raman bands at 493 and 605 cm^{-1}, which were assigned to D_2 and D_1 defect modes of three and four siloxane rings of SBA-15 support material. [111, 122, 123] Another band at 977 cm^{-1} corresponded to vibrations of the Si-O-Si bond in the presence of iron species in the silica framework or defect sites. [111,

124–127] Raman spectrum of SBA-15 support material did not exhibit an analogous Raman band at 977 cm^{-1}. Hence, this Raman band was indicative of pronounced interactions between iron oxidic species and SBA-15 support material.

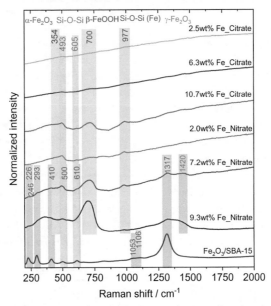

Figure 5.15: Raman spectra of Fe$_x$O$_y$/SBA-15 samples at various iron loadings and mechanical mixture Fe$_2$O$_3$/SBA-15. Characteristic Raman bands are highlighted.

Independent of the precursor, increasing iron loading correlated with increasing intensity of Raman bands. At iron loadings lower than 7.2wt%, all Fe$_x$O$_y$/SBA-15 samples exhibited only characteristic SBA-15 Raman bands, i.e. bands assigned either to Si-O-Si or Si-O-Si next to framework Fe vibrations. The lack of further Raman bands was ascribed to the high dispersion of the small iron oxidic species supported on SBA-15. Bands in the Raman spectra of higher loaded Fe$_x$O$_y$/SBA-15 samples agreed with the increased iron oxidic species size, being consistent with the results from DR-UV-Vis spectroscopy. Intensity of the Raman bands of citrate samples was significantly lower than that of corresponding nitrate samples, indicating a precursor-dependent particle size effect. Consequently, as already revealed from DR-UV-Vis spectroscopy, the citrate precursor induced smaller, higher dispersed iron oxidic species compared to the nitrate precursor.

For gaining insight into differences in iron oxidic species dependent on the precursor, Raman spectra of highest loaded citrate and nitrate samples were further analyzed and compared with Fe$_2$O$_3$/SBA-15 (Figure 5.16). For the mechanical mixture, Fe$_2$O$_3$/SBA-15, main Raman bands at 226, 246, 293, 410, 500, 613, 1053, and 1106 cm^{-1} were observed, corresponding to α-Fe$_2$O$_3$. [120, 121, 123] Intensities of the observed bands in the Raman spectra of the mechanical mixture were high compared to those of the Fe_xO_y/SBA-15 samples. This agreed with the presence of crystalline α-Fe_2O_3 being mixed with SBA-15 in the mechanical mixture. For 9.3wt% Fe_Nitrate, characteristic Raman bands of α-Fe_2O_3 were also revealed. However, intensity of these bands was very weak, indicating a high dispersion of small iron oxidic species on SBA-15. Since a laser power of < 0.5 mW was applied for Raman spectra acquisition, characteristic Raman bands of α-Fe_2O_3, originating from a laser-induced oxidation of other iron species, as reported by Faria et al. [120] and Shebanova and Lazor [128], were excluded. Furthermore, broad and comparatively intense Raman bands around 354 and 700 cm^{-1} were observed for 9.3wt% Fe_Nitrate, which were assigned to β-FeOOH. Measurements of an extended Raman shift range (Figure 5.15) revealed a broad shoulder around 1420 cm^{-1} for 9.3wt% Fe_Nitrate and also for 7.2wt% Fe_Nitrate. This shoulder may correspond to γ-Fe_2O_3. [120, 129] However, all other characteristic features of γ-Fe_2O_3 in Raman shift range between 640 and 725 cm^{-1} were not observed. Therefore, presence of γ-Fe_2O_3 in Fe_xO_y/SBA-15 samples seemed rather improbable. Conversely, for 10.7wt% Fe_Citrate, only Raman bands assigned to β-FeOOH and none assigned to α-Fe_2O_3 were observed. Hence, Raman spectroscopy revealed significant precursor-dependent differences in iron oxidic species. Nitrate precursor presumably induced two different iron oxidic species, α-Fe_2O_3-like and β-FeOOH-like species. Conversely, the citrate precursor yielded at least β-FeOOH-like species, although presence of additional α-Fe_2O_3-like species could not be excluded due to the weak Raman bands.

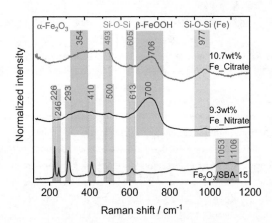

Figure 5.16: Raman spectra of highest loaded citrate sample (top), highest loaded nitrate sample (middle), and mechanical mixture Fe_2O_3/SBA-15 (bottom).

Mössbauer spectroscopy

The chemical state of iron in supported iron oxidic species on SBA-15 was additionally investigated by Mössbauer spectroscopy. Mössbauer spectra of highest loaded Fe_xO_y/SBA-15 samples, i.e. 9.3wt% Fe_Nitrate and 10.7wt% Fe_Citrate, recorded above 200 K showed a broadened and asymmetric doublet independent of the precursor. Therefore, this doublet was analyzed using two non-equivalent Fe sites. The determined values for the isomer shift, δ, and the quadrupole shift, ε, were consistent with those reported for superparamagnetic particles of α-Fe$_2$O$_3$. [39, 50] At low temperatures, that is, at 14 K for 9.3wt% Fe_Nitrate and at 4 K for 10.7wt% Fe_Citrate (Figure 5.17, left), the doublet almost disappeared and a magnetically split hyperfine pattern was detected. This observation indicated the presence of small superparamagnetic iron oxidic species. The related Mössbauer parameters, summarized in Table 5.3, were furthermore consistent with those reported for (magnetically blocked) superparamagnetic particles with a local geometry similar to that of α-Fe$_2$O$_3$ supported on SBA-15. [39] Therefore, a blocking temperature lower than 200 K implied an upper limit for the iron oxidic species diameter of < 10 nm [130] for 9.3wt% Fe_Nitrate. Conversely, observing an almost complete blocking at lower temperatures for 10.7wt% Fe_Citrate compared to 9.3wt% Fe_Nitrate suggested a significantly smaller iron oxidic species size obtained from citrate precursor. Furthermore, refinement of the Mössbauer spectra of 9.3wt% Fe_Nitrate and 7.2wt% Fe_Nitrate at 14 K and of 10.7wt% Fe_Citrate at 14 K and 4 K required an additional component indicating a bimodal particle size distribution (Table 5.3). For these low temperature Mössbauer spectra, only combination of one Fe site for the doublet and two non-equivalent Fe sites for the sextet yielded satisfying fit results. Determined values of the local magnetic hyperfine field, B_{hf}, for the required two Fe sites for the sextet differed, and hence, confirmed a bimodal particle size distribution. Lower B_{hf} values could be ascribed to smaller Fe$_x$O$_y$ species and higher B_{hf} values to larger Fe$_x$O$_y$ species, respectively.

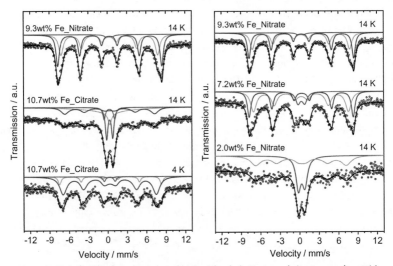

Figure 5.17: Left: Mössbauer spectra of highest loaded nitrate and citrate samples at 14 and 4 K. Right: Mössbauer spectra of nitrate samples at 14 K with various iron loadings.

The Mössbauer spectra of lower loaded nitrate samples, 7.2wt% Fe_Nitrate and 2.0wt% Fe_Nitrate, also exhibited a broadened and asymmetric doublet at 300 K with similar values for isomer shift and quadrupole shift as determined for 9.3wt% Fe_Nitrate. While this doublet almost completely transformed into a magnetically split sextet for sample 9.3wt% Fe_Nitrate at 14 K (vide supra), this transformation remained incomplete in the Mössbauer spectra of 7.2wt% Fe_Nitrate and 2.0wt% Fe_Nitrate at 14 K (Figure 5.17, right). The site population ratio of the doublet relative to the magnetically split sextet signal decreased systematically with decreasing iron loading from about 2:98 for 9.3wt% Fe_Nitrate to 45:55 for 2.0wt% Fe_Nitrate (Table 5.3). Furthermore, a similar trend was observed for the local magnetic hyperfine field, i.e. decreasing B_{hf} with decreasing iron loading. Assuming that all iron species in the nitrate samples consisted of mainly iron oxide, both results independently suggested a correlation of increasing average iron oxidic species size and increasing iron loading within the nitrate samples.

Refinement of the Mössbauer spectra of 10.7wt% Fe_Citrate at low temperatures still required two non-equivalent Fe sites for the doublet, while for 9.3wt% Fe_Nitrate, two non-equivalent Fe sites were no longer necessary for sufficient fit results. This was possibly indicative of two different types of iron oxidic species on SBA-15 dependent on used precursor. Based on this observation, fitted doublet in low temperature Mössbauer spectra was more precisely analyzed. Crystalline β-FeOOH is paramagnetic at 300 K and magnetically ordered at 77 K. [131] However, assuming the transformation of the doublet into a magnetically split sextet for β-FeOOH being analogously dependent on particle size as for α-Fe$_2$O$_3$, Mössbauer parameters of fitted doublet at 14 K might be indicative of β-FeOOH-like species. Values of isomer shift and quadrupole shift at 14 K matched those reported for paramagnetic β-FeOOH. This might be ascribed to small and highly dispersed β-FeOOH-like species in Fe$_x$O$_y$/SBA-15 samples. Area from refinement of the doublet could be correlated with the amount of β-FeOOH-like species. However, low iron content of Fe$_x$O$_y$/SBA-15 samples induced a low statistical certainty. Therefore, only quantitative estimation of the phase fractions was meaningful. Apparently, the amount of β-FeOOH-like species increased within the nitrate samples with decreasing iron loading and the citrate precursor induced a higher amount of β-FeOOH-like species compared to the nitrate precursor.

Information on coordination environment of the iron centers might be deduced from values of isomer shift. Samanta et al. [132] reported a correlation between isomer shift of room temperature spectra and coordination geometry of iron in Fe-MCM-41 samples. Assuming this correlation being applicable to Fe$_x$O$_y$/SBA-15 samples, average isomer shift values higher than 0.3 mm/s for Mössbauer spectra of all samples at ambient temperature indicated the presence of octahedrally coordinated Fe^{3+} species (Table 5.3).

Table 5.3: Mössbauer parameters for 9.3wt% Fe_Nitrate, 7.2wt% Fe_Nitrate, 2.0wt% Fe_Nitrate, and 10.7wt% Fe_Citrate. Temperature, T, isomer shift, δ (referred to α-Fe at 298 K and not corrected for 2nd order Doppler shift), quadrupole shift, ε, line widths, Γ_{HWHM}, hyperfine magnetic field, B_{hf}, fluctuation rate, v_c, and area. [*] values held fixed in simulation. [a] relaxation rate reached the dynamic limit.

Sample	T / K	δ / mm/s	ε / mm/s	Γ_HWHM / mm/s	B_hf / T	v_c / mm/s	Area / %
9.3wt% Fe_Nitrate	300	0.320(9)	0.173(42)	0.29(11)	48.3*	[a]	48
		0.327(8)	0.536(37)	0.277(45)	48.3*	[a]	52
	14	0.401(21)	-0.012(20)	0.28*	46.5(2.7)	0.13	45
		0.465(11)	-0.018(11)	0.28*	50.0(1.2)	0.02	52
		0.401*	0.43*	0.37*	48.3*	310*	3
7.2wt% Fe_Nitrate	300	0.330(5)	0.299(15)	0.190(16)	48.3*	[a]	52
		0.307(7)	0.508(27)	0.233(17)	48.3*	[a]	48
	14	0.394(35)	-0.014(43)	0.23*	45.4(5)	0.3	49
		0.462(19)	-0.034(19)	0.23*	49.7(2)	0.1	39
		0.431(72)	0.518(59)	0.45*	48.3*	[a]	12
2.0wt% Fe_Nitrate	300	0.336(15)	0.346(78)	0.273(60)	48.3*	[a]	60
		0.312(21)	0.583(95)	0.265(71)	48.3*	[a]	40
	14	0.421*	0.08(12)	0.24(76)	43.8(1.5)	0.7	55
		0.423(48)	0.500(60)	0.437(80)	48.3*	520	45
10.7wt% Fe_Citrate	300	0.294(12)	0.206(39)	0.34(12)	48.3*	[a]	45
		0.316(12)	0.672(50)	0.376(49)	48.3*	[a]	55
	14	0.451*	-0.008(64)	0.20*	43.5*	5.6	34
		0.451(10)	-0.005(97)	0.20*	43.5(7)	0.5	31
		0.438(15)	0.466(22)	0.23*	48.3*	[a]	25
		0.416(38)	0.814(50)	0.23*	48.3*	[a]	10
	4	0.497(62)	0.018(62)	0.20*	48.9(5)	0.05	14
		0.424(47)	-0.026(45)	0.20*	45.0(6)	0.45	81
		0.438*	0.47*	0.23*	48.3*	[a]	1
		0.416*	0.81*	0.23*	48.3*	[a]	4

Combination of results from DR-UV-Vis, Raman, and Mössbauer spectroscopy

Based on only one of these spectroscopic methods, meaningful characterization of Fe_xO_y/SBA-15 samples was rather difficult, especially at low iron loadings. However, revealed effects or results corroborated by at least two of these spectroscopic methods were well-founded. Overview of results of applied spectroscopic methods for characterization of Fe_xO_y/SBA-15 samples is summarized in Table 5.4. A particle size effect dependent on iron loading was revealed by DR-UV-Vis, Raman, and Mössbauer spectroscopy. Additionally, precursor-dependent particle size effect, i.e. smaller iron oxidic species obtained from citrate precursor, was corroborated by all three methods. Besides particle size effects, differences in types of iron oxidic species were observed. DR-UV-Vis spectra of 10.7wt% Fe_Citrate, 9.3wt% Fe_Nitrate, and 7.2wt% Fe_Nitrate showed absorption bands characteristic of small α-Fe$_2$O$_3$-like species. However, further characteristic absorption bands corresponding to small Fe$_x$O$_y$ oligomers and isolated Fe^{3+} species indicated only a small fraction of α-Fe$_2$O$_3$-like species in the Fe$_x$O$_y$/SBA-15 samples. This was in accordance with the observed Raman bands assigned to α-Fe$_2$O$_3$ for the nitrate samples. Furthermore, Mössbauer parameters agreed with the presence of small α-Fe$_2$O$_3$-like species. An average isomer shift higher than 0.3 mm/s in Mössbauer spectra recorded at ambient temperature indicated octahedrally coordinated Fe^{3+} species. This complied with characteristic absorption bands of CT transitions of both isolated octahedrally coordinated Fe^{3+} and octahedrally coordinated Fe^{3+} in small Fe$_x$O$_y$ oligomers, as revealed by DR-UV-Vis spectroscopy. A possible bimodal particle size distribution was also corroborated by Mössbauer and DR-UV-Vis spectroscopy. The probably most important result revealed by Raman and Mössbauer spectroscopy was a precursor-dependent difference in types of iron oxidic species. Apparently, the citrate precursor yielded a significant amount of β-FeOOH-like species on SBA-15, whereas the nitrate precursor induced both, α-Fe$_2$O$_3$-like and β-FeOOH-like species. Differences in types of iron oxidic species were furthermore in accordance with differences in sample color (Figure 5.2). To further elucidate the influence of the precursor on resulting iron oxidic species, especially on local structure of Fe$_x$O$_y$/SBA-15 samples, X-ray absorption spectroscopy was applied (see chapter 5.3.4)

Table 5.4: Overview of results of applied spectroscopic methods for characterization of Fe$_x$O$_y$/SBA-15 samples. ✔: method corroborated result. ✘: no conclusion possible.

	DR-UV-Vis	Raman	Mössbauer
Particle size effect dependent on iron loading	✔	✔	✔
Precursor-dependent particle size effect	✔	✔	✔
Fraction of small α-Fe$_2$O$_3$-like species	✔	✔	✔
Octahedrally coordinated Fe^{3+} species	✔	✘	✔
Bimodal particle size distribution	✔	✘	✔
Precursor-dependent differences in iron oxidic species (i.e. β-FeOOH- and α-Fe$_2$O$_3$-like)	✘	✔	✔

5.3.4 Local structure of Fe$_x$O$_y$/SBA-15

X-ray absorption near edge structure

Fe K edge X-ray absorption near edge structure (XANES) spectra of various iron oxide references, mechanical mixture, and Fe$_x$O$_y$/SBA-15 samples at highest, middle, and lowest iron loadings are depicted in Figure 5.18. All Fe$_x$O$_y$/SBA-15 samples exhibited a similar overall shape of XANES spectra. However, slight differences in XANES spectra of Fe$_x$O$_y$/SBA-15 samples dependent on the precursor were discernible. Citrate samples showed a broader absorption in XANES region beyond the absorption edge, compared to the nitrate samples. This might be indicative of precursor-dependent differences in local structure of resulting iron oxidic species. Comparing Fe$_x$O$_y$/SBA-15 XANES spectra with iron oxide reference spectra indicated similarities to α-Fe_2O_3 reference spectrum for nitrate samples and to β-FeOOH reference spectrum for citrate samples. XANES spectrum of the mechanical mixture corresponded to α-Fe_2O_3 reference spectrum, as expected for α-Fe_2O_3 particles being mixed with SBA-15.

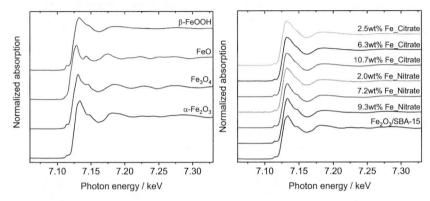

Figure 5.18: Fe K edge XANES spectra of various iron oxide references (left), Fe_xO_y/SBA-15 samples at various iron loadings (right), and mechanical mixture Fe_2O_3/SBA-15 (right).

All Fe_xO_y/SBA-15 XANES spectra exhibited a small pre-edge peak resulting from dipole forbidden $1s \rightarrow 3d$ electron transition. The pre-edge peak gains intensity, on the one hand, through deviation from centrosymmetric coordination, i.e. distortions from perfect octahedral coordination symmetry. On the other hand, through 3p-4d orbital mixing, i.e. p-d hybridization whereby the final state obtains p-character. A sharp and intense pre-edge peak is commonly ascribed to the presence of tetrahedrally coordinated species without inversion center. [133–135] Therefore, weak pre-edge peaks in XANES spectra of Fe_xO_y/SBA-15 samples implied slightly distorted octahedral coordination of supported iron oxidic species independent of both iron loading and precursor. Refining two Pseudo-Voigt functions to the pre-edge peak feature in experimental XANES spectra of all Fe_xO_y/SBA-15 samples afforded a more detailed analysis.

Pre-edge peak features are often used to investigate changes in coordination geometry or oxidation state. Commonly, pre-edge peak height is applied for a quantitative analysis of the coordination symmetry. [133–136] Besides analysis of coordination symmetry, pre-edge peak feature can also be used for estimating the reducibility of Fe^{3+}. Petit et al. [137] estimated Fe^{3+}/Fe^{2+} ratio in minerals based on the pre-edge peak area. Pre-edge peak heights of Fe_xO_y/SBA-15 ranged from 0.05 through 0.06 for the citrate samples, and from 0.06 through 0.07 for the nitrate samples (Figure 5.19, left). These similar and low pre-edge peak heights were indicative of a similar octahedral coordination geometry of all Fe_xO_y/SBA-15 samples. However, pre-edge peak area as function of iron loading showed slight differences dependent on the precursor (Figure 5.19, right). An opposite behavior for citrate and nitrate samples was revealed. Pre-edge peak area decreased with increasing iron loading for citrate samples, indicative of a decreasing reducibility with increasing iron loading. Conversely, increasing iron loading within the nitrate samples correlated with an increasing pre-edge peak area, and hence, an increasing reducibility.

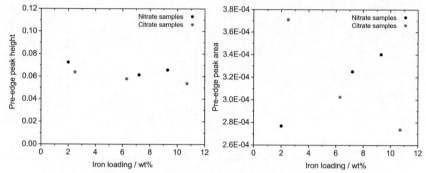

Figure 5.19: Pre-edge peak height (left) and pre-edge peak area (right) as function of iron loading for Fe_xO_y/SBA-15 samples. Values were obtained from pre-edge peak fitting using two Pseudo-Voigt functions.

Fe K edge XANES spectra were furthermore used for determining the average oxidation state of iron centers of Fe_xO_y/SBA-15 samples. Therefore, a least-squares fit was carried out to the first absorption peak in the pre-edge region of XANES spectra to simulate the absorption edge (Figure A.12.2). The centroid of the fitted arctangent (arctan) function represents the absorption edge. Hence, position of centroid of the fitted arctangent function was used for determining the average iron oxidation state. Therefore, various iron oxide references were used for constructing a linear calibration (Figure 5.20). Linear calibration could be expressed by:

Average Fe oxidation state = 205.22 ($\pm$ 18.88) keV^{-1} $\cdot$ Pos $-$ 1458.91 ($\pm$ 134.45)   (5.5)

with absorption edge position, *Pos*. Linear correlation between position of absorption edge and average iron oxidation state was directly applied to Fe_xO_y/SBA-15 samples yielding an average iron oxidation state of +3 for all Fe_xO_y/SBA-15 samples (Table 5.5). However, Table 5.5 shows a slight deviation of average iron oxidation state for lowest loaded citrate sample, 2.5wt% Fe_Citrate. This deviation was attributed to the use of crystalline iron oxide references for linear calibration. Especially for lowest loaded samples, crystalline iron oxide references differed from supported iron oxides in dispersion of iron centers and local geometry. These properties influence not only the position of Fe K edge, which was used for calibration, but also the average oxidation state. However, assuming fully oxidized iron oxidic species on SBA-15 after calcination agreed with obtained values of average iron oxidation state.

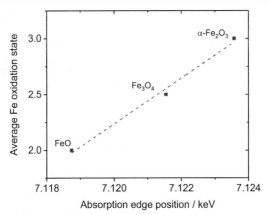

Figure 5.20: Linear relationship between absorption edge position and average iron oxidation state for various iron oxide references.

Table 5.5: Average iron oxidation state of Fe$_x$O$_y$/SBA-15 samples calculated from linear calibration (eq.(5.5)).

	Average Fe oxidation state		Average Fe oxidation state
2.5wt% Fe_Citrate	2.8	2.0wt% Fe_Nitrate	3.0
6.3wt% Fe_Citrate	3.0	7.2wt% Fe_Nitrate	3.0
10.7wt% Fe_Citrate	3.0	9.3wt% Fe_Nitrate	3.0

Additionally, for determining phase composition of Fe$_x$O$_y$/SBA-15 samples, experimental XANES spectra were simulated by fitting various linear combinations (LC) of iron oxide reference spectra. For all Fe$_x$O$_y$/SBA-15 samples, only combination of two different iron oxide reference spectra yielded satisfying fit results. Required iron oxide references were α-Fe$_2$O$_3$ and β-FeOOH. Hence, Fe$_x$O$_y$/SBA-15 samples consisted of iron oxidic species with a local geometry similar to that of α-Fe$_2$O$_3$ or β-FeOOH. Figure 5.21 depicts the refinement results of the sum of the reference XANES spectra to experimental Fe$_x$O$_y$/SBA-15 XANES spectra. Independent of the precursor, the amount of β-FeOOH-like species increased with decreasing iron loading. Lowest loaded citrate sample possessed 25.2% of β-FeOOH-like species, while fraction of β-FeOOH-like species for lowest loaded nitrate sample was determined to be 15.3% (Table 5.6). Hence, pronounced differences in phase composition of Fe$_x$O$_y$/SBA-15 samples were observed dependent on precursor. The citrate samples possessed significantly higher amounts of β-FeOOH-like species compared to corresponding nitrate samples with similar iron loading.

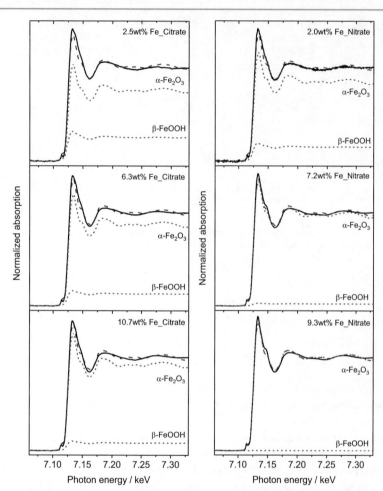

Figure 5.21: Refinement of sum (dashed lines) of α-Fe$_2$O$_3$ and β-FeOOH reference (dotted lines) to experimental Fe K edge XANES spectra (straight lines) of Fe$_x$O$_y$/SBA-15 samples with increasing iron loading from top to bottom. Left: Citrate samples. Right: Nitrate samples.

Table 5.6: Fraction of α-Fe$_2$O$_3$- and β-FeOOH-like species in Fe$_x$O$_y$/SBA-15 obtained from phase composition analysis of Fe K edge XANES spectra. Corresponding refinements are shown in Figure 5.21.

	α-Fe$_2$O$_3$ / %	β-FeOOH / %		α-Fe$_2$O$_3$ / %	β-FeOOH / %
2.5wt% Fe_Citrate	74.8	25.2	2.0wt% Fe_Nitrate	84.7	15.3
6.3wt% Fe_Citrate	87.0	13.0	7.2wt% Fe_Nitrate	97.6	2.4
10.7wt% Fe_Citrate	91.8	8.2	9.3wt% Fe_Nitrate	99.9	0.1

Extended X-ray absorption fine structure

Extended X-ray absorption fine structure (EXAFS) analysis of the Fe$_x$O$_y$/SBA-15 samples at highest, middle, and lowest iron loading yielded additional information on local structure of supported iron centers. The pseudo radial distribution function (not phase shift corrected), FT(χ(k)*k^3), of all Fe$_x$O$_y$/SBA-15 samples exhibited two prominent peaks at a distance around 1.5 Å and around 2.7 Å, respectively. The first peak in the pseudo radial distribution function at ~ 1.5 Å corresponded to Fe-O shells, whereas the second peak at ~ 2.7 Å corresponded to Fe-Fe shells. The peak positions in the FT(χ(k)*k^3) of the Fe$_x$O$_y$/SBA-15 samples were in good agreement with those in the reference spectrum of α-Fe$_2$O$_3$. Figure 5.22 depicts the pseudo radial distribution function of all Fe$_x$O$_y$/SBA-15 samples and that of α-Fe$_2$O$_3$ reference. The similar shape of the FT(χ(k)*k^3) of the Fe$_x$O$_y$/SBA-15 samples and the reference α-Fe$_2$O$_3$ indicated a similar local structure around the iron centers in the Fe$_x$O$_y$/SBA-15 samples and α-Fe$_2$O$_3$, and hence, justified EXAFS refinements using α-Fe$_2$O$_3$ as structural model.

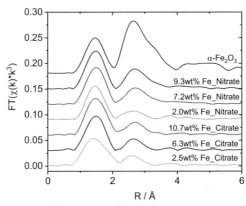

Figure 5.22: Pseudo radial distribution function (not phase shift corrected), FT(χ(k)*k^3), of α-Fe$_2$O$_3$ reference and Fe$_x$O$_y$/SBA-15 samples with an offset in y-axis for clarity.

Figure 5.23 depicts the X-ray absorption fine structure (XAFS), χ(k)*k^3 (left), together with the pseudo radial distribution function, FT(χ(k)*k^3) (right), of citrate (top) and nitrate (bottom) samples. Data quality of the X-ray absorption fine structure permitted reliable data analysis up to 10.3 Å^{-1} except for sample 2.5wt% Fe_Citrate, where reliable data analysis was merely permitted up to 9.1 Å^{-1} (Figure 5.23, left). The FT(χ(k)*k^3) of the Fe$_x$O$_y$/SBA-15 samples showed significant differences dependent on the precursor (Figure 5.23, right). The amplitude of both first and second peak in the FT(χ(k)*k^3) of the citrate samples was lower compared to that of corresponding nitrate samples. Furthermore, within the citrate samples, increasing iron loading induced a slight increase in amplitude of first and second peak in the FT(χ(k)*k^3). However, this influence of iron loading for the citrate samples was not considered as significant effect due to poor data quality of the χ(k)*k^3. Within the nitrate samples, amplitude of the first peak in the FT(χ(k)*k^3) also showed a slight increase in intensity with increasing iron loading, while that of the second peak was more pronounced. As for the citrate samples, influence of iron loading on amplitude of the first peak in the FT(χ(k)*k^3), assigned to Fe-O shells, was considered negligible. Conversely, influence of iron loading on amplitude of the second peak in the FT(χ(k)*k^3), corresponding to Fe-Fe shells, was pronounced for the nitrate samples. Based on these observations, amplitude of the first peak in the FT(χ(k)*k^3) of Fe$_x$O$_y$/SBA-15 samples, was considered rather independent of iron loading. However, amplitude of the second peak in the FT(χ(k)*k^3) showed an iron loading dependence, and moreover, a precursor dependence.

EXAFS refinements for the Fe$_x$O$_y$/SBA-15 samples were performed to further elucidate the local structure around the supported iron centers dependent on both iron loading and precursor. Therefore, theoretical X-ray absorption fine structure calculated for a α-Fe$_2$O$_3$ model structure, consisting of corner-, edge-, and face-shared [FeO$_6$] octahedra, was refined to the experimental FT(χ(k)*k^3) of Fe$_x$O$_y$/SBA-15 samples. To ensure statistical significance of the parameters during the fitting procedure, the number of free running parameters, N_{free}, was kept well below the number of independent parameters, N_{ind}, according to the Nyquist theorem. Furthermore, the number of considered backscattering paths was kept as low as possible. For each refinement, confidence limits and so-called F-Tests were calculated to assure statistical certainty of the fitting parameters. Fit results are summarized in Table 5.7. Within the refinements, the average coordination number, N, of each backscattering path was kept invariant and the E_0 shifts of all required backscattering paths were correlated to be equal. A good agreement between theoretical and experimental FT(χ(k)*k^3) of the Fe$_x$O$_y$/SBA-15 samples was achieved using α-Fe$_2$O$_3$ as structural model. Theoretical and experimental FT(χ(k)*k^3) are shown in Figure 5.24. Within the limits of distance resolution in EXAFS refinements (dR = π (2 k_{max})$^{-1}$ = 0.15 Å for k_{max} = 10.3 Å^{-1} (Table 5.7)) only one Fe-O distance was resolved. Dependent on the precursor, significant differences in EXAFS refinements were revealed. For the citrate samples, amplitude in the FT(χ(k)*k^3) was fully accounted for by using one Fe-O and one Fe-Fe distance within the refinement. For lowest loaded nitrate sample, 2.0wt% Fe_Nitrate, an analogous fitting procedure yielded good fit results. Conversely, for nitrate samples with iron loadings higher than 2.0wt%, an additional Fe-Fe distance was required. The contribution of a higher Fe-Fe distance for higher loaded nitrate samples was in good agreement with DR-UV-Vis and Mössbauer spectroscopy results. This confirmed an increasing Fe_xO_y species size by using the nitrate precursor, and furthermore, with increasing iron loading.

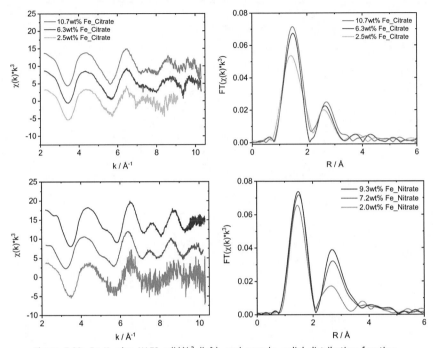

Figure 5.23: Fe K edge XAFS, $\chi(k)*k^3$ (left), and pseudo radial distribution function, $FT(\chi(k)*k^3)$ (right), of citrate samples (top) and nitrate samples (bottom).

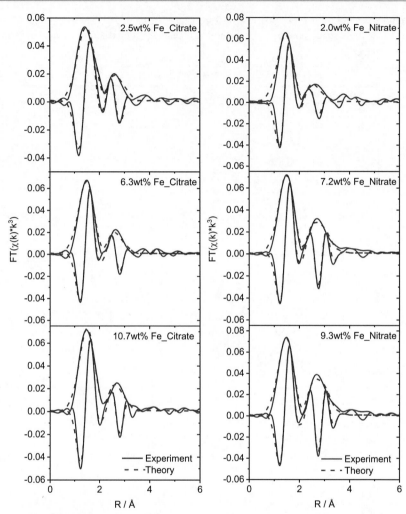

Figure 5.24: Experimental (straight line) and theoretical (dotted line) Fe K edge FT(χ(k)*k^3) of Fe$_x$O$_y$/SBA-15 samples.

Table 5.7: EXAFS fit results for Fe_xO_y/SBA-15 samples and α-Fe_2O_3. Type of neighbor, coordination number, N, and XAFS disorder parameter, σ^2, of atoms at the distance R from the Fe atoms in Fe_xO_y/SBA-15 samples at various iron loadings. Experimental parameters were obtained from refinement of α-Fe_2O_3 model structure (AMCSD 0017806 [138]) to the experimental Fe K edge FT(χ(k)*k^3) of Fe_xO_y/SBA-15 (Figure 5.24). Fit range in R space 1.1-3.4 Å (or 1.1-3.0 Å); k space 2.2-10.3 Å^{-1} (or 2.2-9.1 Å^{-1}); N_{ind} = 13.4 (or 11.3); N_{free} = 7 (or 5); Number of SS paths = 3 (or 2); S_0^2 = 0.9. Alternative values of fitting parameters given in parentheses refer to slightly modified refinement for sample 2.5wt% Fe_Citrate due to χ(k)*k^3 data quality.

	Type	N$_{fixed}$	R / Å	σ^2 / Å^2	E$_0$ / eV	Fit residual R
2.5wt% Fe_Citrate	Fe-O	6	1.96	0.0187	+3.24	7.0
	Fe-Fe	3	3.04	0.0169		
	Fe-Fe	3	-	-		
6.3wt% Fe_Citrate	Fe-O	6	1.95	0.0147	+1.99	7.3
	Fe-Fe	3	3.04	0.0177		
	Fe-Fe	3	-	-		
10.7wt% Fe_Citrate	Fe-O	6	1.97	0.0135	+3.16	5.9
	Fe-Fe	3	3.04	0.0158		
	Fe-Fe	3	-	-		
2.0wt% Fe_Nitrate	Fe-O	6	1.94	0.0157	+0.95	6.5
	Fe-Fe	3	2.99	0.0200		
	Fe-Fe	3	-	-		
7.2wt% Fe_Nitrate	Fe-O	6	1.95	0.0138	+2.83	6.2
	Fe-Fe	3	3.01	0.0154		
	Fe-Fe	3	3.42	0.0178		
9.3wt% Fe_Nitrate	Fe-O	6	1.94	0.0134	+2.01	6.9
	Fe-Fe	3	3.01	0.0136		
	Fe-Fe	3	3.41	0.0147		
α-Fe_2O_3	Fe-O	6	1.96	0.0131	-0.7	8.1
	Fe-Fe	3	2.94	0.0055		
	Fe-Fe	3	3.30	0.0021		

For all Fe_xO_y/SBA-15 samples, the coordination number for the Fe-O backscattering path was kept invariant at six due to the presence of mostly [FeO$_6$] octahedral units in α-Fe_2O_3 model structure. Furthermore, XANES, DR-UV-Vis, and Mössbauer spectroscopy analysis did not indicate the presence of tetrahedrally coordinated iron oxidic species. Considering a phase fraction of β-FeOOH probably being present in Fe_xO_y/SBA-15 samples, as suggested in chapter 5.3.3 and by XANES analysis, an invariant coordination number of six for the Fe-O backscattering path was still meaningful. Structure of β-FeOOH also consists of [FeO$_6$] octahedral units. [139] Hence, variations in the amplitude in the FT(χ(k)*k^3) of Fe_xO_y/SBA-15 samples were not induced by variations in the coordination number. The static disorder in the phase fraction of β-FeOOH and α-Fe$_2$O$_3$ was assumed to be largely independent of the iron loading. Consequently, only the variations in amount of the phase fractions were dependent on iron loading. Accordingly, the local structure within the phase fractions was largely independent of iron loading, and hence, did not affect the Fe-O shell. Despite, Fe$_x$O$_y$ species size was dependent on iron loading (chapter 5.3.3). However, at these low iron loadings, all Fe$_x$O$_y$ species were significantly smaller than crystalline β-FeOOH and α-Fe$_2$O$_3$ and the local structure was presumably not influenced by varying species size. Variations in the amplitude in the FT(χ(k)*k^3) appearing in the Debye-Waller factors, σ^2(Fe-O), were thus induced by variations in the phase composition (Figure 5.25, left). All Fe$_x$O$_y$/SBA-15 samples showed a decreasing σ^2(Fe-O) with increasing iron loading. And σ^2(Fe-O) of the citrate samples were higher than those of corresponding nitrate samples. XANES analysis revealed the presence of β-FeOOH-like and α-Fe$_2$O$_3$-like species in the Fe$_x$O$_y$/SBA-15 samples and a decreasing amount of β-FeOOH-like species with increasing iron loading independent of the precursor. Accordingly, σ^2(Fe-O) might be assigned to the phase fraction of β-FeOOH-like species, and hence, to the presence of Fe-O and Fe-OH bonds in Fe$_x$O$_y$/SBA-15 samples. The citrate samples possessed a higher amount of β-FeOOH-like species being correlated with the higher values of σ^2(Fe-O) compared to corresponding nitrate samples. Not only for the citrate but also for the nitrate samples, decreasing σ^2(Fe-O) with increasing iron loading was assigned to decreasing amount of β-FeOOH-like species (Figure 5.25, left and Table 5.6). Moreover, the slightly longer Fe-O distances of the citrate samples, compared to those of corresponding nitrate samples, might also be ascribed to different amounts of β-FeOOH-like species. Fe-OH bonds possess longer distances than Fe-O bonds. As a result of this difference in bond distance, the higher amount of β-FeOOH-like species in Fe$_x$O$_y$/SBA-15 samples obtained from citrate precursor might become visible in the slightly higher values of R(Fe-O) (Figure 5.25, right). Nevertheless, considering limits in distance resolution, variations in R(Fe-O) were only minor.

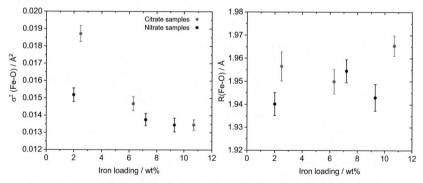

Figure 5.25: Fe-O Debye-Waller factor (XAFS disorder parameter), σ^2 (left), and Fe-O distance, R (right), of Fe$_x$O$_y$/SBA-15 samples as function of iron loading. Parameters resulted from refinements of Fe K edge FT(χ(k)*k^3) (Table 5.7).

Significant differences dependent on the precursor were also revealed in Debye-Waller factor, $\sigma^2(Fe\text{-}Fe\_edge)$, and Fe-Fe distance, $R(Fe\text{-}Fe\_edge)$, of edge-shared [FeO$_6$] octahedra of Fe$_x$O$_y$/SBA-15 samples (Figure 5.26). For variations in $\sigma^2(Fe\text{-}Fe\_edge)$ and $R(Fe\text{-}Fe\_edge)$, the influence of average coordination number was considered highly relevant, while that of the phase composition, i.e. amount of β-FeOOH-like and α-Fe$_2$O$_3$-like species, was considered minor. Hence, assuming the amount of β-FeOOH-like species slightly affected $\sigma^2(Fe\text{-}Fe\_edge)$ and $R(Fe\text{-}Fe\_edge)$, two superimposed effects had to be considered for the Fe-Fe shell. All Fe$_x$O$_y$/SBA-15 samples showed a decreasing $\sigma^2(Fe\text{-}Fe\_edge)$ with increasing iron loading (Figure 5.26, left). This decrease in $\sigma^2(Fe\text{-}Fe\_edge)$ was more pronounced for the nitrate samples. The average coordination number for the Fe-Fe backscattering path was set to three and kept invariant within the EXAFS refinements. However, at low iron loadings, the average coordination number was presumably lower than three. Therefore, decreasing iron loading induced a decreasing amplitude in the FT(χ(k)*k^3), and hence, an increasing $\sigma^2(Fe\text{-}Fe\_edge)$.

All citrate samples possessed higher $\sigma^2(Fe\text{-}Fe\_edge)$ compared to corresponding nitrate samples, except for 2.5wt% Fe_Citrate. This was correlated with smaller Fe$_x$O$_y$ species sizes and lower degree of aggregation yielding a decrease in average coordination number of citrate samples, compared to corresponding nitrate samples. This correlation agreed well with the precursor-dependent particle size effect of Fe$_x$O$_y$/SBA-15 samples (see chapter 5.3.3). Edge-shared [FeO$_6$] octahedral units are present in crystalline β-FeOOH and α-Fe$_2$O$_3$. Theoretical distance of Fe-Fe bonds in edge-shared octahedra of crystalline β-FeOOH possesses a value of 3.04 Å, whereas that of crystalline α-Fe$_2$O$_3$ is calculated to be 2.97 Å. [139] This difference in theoretical Fe-Fe distance of edge-shared octahedra in crystalline β-FeOOH and α-Fe$_2$O$_3$ seemed to be responsible for the differences in $R(Fe\text{-}Fe\_edge)$ of Fe$_x$O$_y$/SBA-15 samples dependent on the precursor (Figure 5.26, right). The higher Fe-Fe distances of the citrate samples compared to those of corresponding nitrate samples were ascribed to higher amounts of β-FeOOH-like species, and thus, to longer $R(Fe\text{-}Fe\_edge)$. Apparently, iron loading did not significantly affect the Fe-Fe distance of edge-shared [FeO$_6$] octahedra.

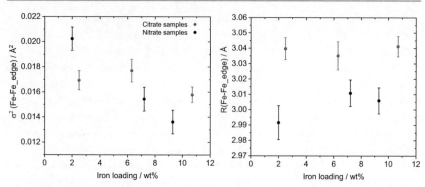

Figure 5.26: Fe-Fe Debye-Waller factor (XAFS disorder parameter), σ^2 (left), and Fe-Fe distance, R (right), of edge-shared octahedra in Fe_xO_y/SBA-15 samples as function of iron loading. Parameters resulted from refinements of Fe K edge FT($\chi(k)*k^3$) (Table 5.7).

In EXAFS refinements of nitrate samples with iron loadings higher than 2.0wt% an additional Fe-Fe backscattering path at higher distances was required to fully account for the amplitude in the FT($\chi(k)*k^3$). Conversely, for citrate samples and lowest loaded nitrate sample, only one Fe-Fe backscattering path was sufficient. The additional Fe-Fe backscattering path at higher distances was ascribed to Fe-Fe in corner-shared [FeO_6] octahedra. Contribution of a higher Fe-Fe distance being only necessary for higher loaded nitrate samples agreed well with largest Fe_xO_y species sizes and highest degree of aggregation for these samples (see chapter 5.3.3). For the Fe-Fe backscattering path ascribed to corner-shared [FeO_6] octahedra, average coordination number was set to three and kept invariant in EXAFS refinements. Accordingly, for higher loaded nitrate samples, average coordination number being lower than three seemed to be responsible for the decrease in the amplitude in the FT($\chi(k)*k^3$). Hence, increasing σ^2(Fe-Fe_corner) from 9.3wt% Fe_Nitrate to 7.2wt% Fe_Nitrate was ascribed to a decreasing average coordination number, and consequently to a decreasing degree of aggregation of Fe_xO_y species (Table 5.7). For R(Fe-Fe_corner), no significant influence of iron loading was revealed (Table 5.7).

Differences in Fe-Fe shells dependent on iron loading and precursor agreed well with both iron loading-dependent and precursor-dependent particle size effect of Fe_xO_y/SBA-15 samples. Furthermore, presence of small and highly dispersed iron oxidic species was confirmed, whereas presence of highly aggregated or even crystalline α-Fe_2O_3-like species could be excluded.

5.3.5 Chemical memory effect of Fe$_x$O$_y$/SBA-15

Previous chapters of structural characterization of Fe$_x$O$_y$/SBA-15 catalysts revealed pronounced differences in Fe$_x$O$_y$ species size, surface and porosity characteristics of Fe$_x$O$_y$/SBA-15, and types of iron oxidic species. Smaller iron oxidic species, accompanied by a smoother support surface and a higher degree of micropore filling, were obtained from the citrate precursor. This was explained by a more pronounced chelating effect of the citrate precursor compared to the nitrate precursor. Conversely, differences in types of iron oxidic species dependent on the precursor were observed but not readily explained. Iron oxidic species supported on SBA-15 possessed a similar geometry to crystal structures of two different space groups, α-Fe$_2$O$_3$ and β-FeOOH. α-Fe$_2$O$_3$ has a corundum-type structure and the space group is $R\bar{3}c$. The unit cell is trigonal and crystal structure exhibits a hexagonal close package. [138, 139] β-FeOOH possesses a cryptomelane-type structure and space group $I4/m$. The unit cell is tetragonal and a hexagonal close package does not exist. [140].

For Cu/ZnO catalysts, a significant influence of the precursor on resulting metal oxide system was reported by several authors and ascribed to the so-called chemical memory effect. [141–143] It was shown that the precursor chemistry controls morphology and defect structure of the final Cu/ZnO system. Apparently, variations in crystallinity, homogeneity, and phase composition of the precursor during synthesis of Cu/ZnO systems are reflected in the structure of Cu/ZnO systems after calcination. [141] A chemical memory effect of amorphous alumosilicate gel precursors was reported by Antonić-Jelić et al. [144]. They observed particular properties of crystallized zeolites being dependent on composition of hydrogel in which the precursor was precipitated. Again, the precursor significantly influenced resulting structure of desired product. Audebrand et al. [145] investigated the influence of the precursor on nanoscale ZnO crystallites. Precursor-dependent differences in microstructural properties of ZnO crystallites could merely be explained by an involved chemical memory effect. Considering the results for Cu/ZnO systems, ZnO crystallites, and zeolites, a chemical memory effect of iron oxidic species on SBA-15 appeared likely. However, results from long-range ordered structures cannot readily be adapted to supported systems. Nevertheless, chemical memory effect of Fe$_x$O$_y$/SBA-15 catalysts might be a possible explanation for the revealed precursor-dependent differences in iron oxidic species.

Again, influence of the different strengths of chelating effect of the two precursors might be involved. The citrate precursor with a more pronounced chelating effect was presumably highly dispersed within the pore system of SBA-15 support material. Formation of hydrogen bonds between citrate precursor and superficial silanol groups of SBA-15, as reported by Gabelica et al. [55] for Fe(III) chelate precursors, seemed likely. Conversely, less hydrogen bonds between silanol groups of SBA-15 and nitrate precursor might result due to a less pronounced chelating effect. These differences in amounts of hydrogen bonds between precursor and SBA-15 support material during synthesis might induce the different amounts of β-FeOOH-like species, assuming a chemical memory of Fe$_x$O$_y$/SBA-15.

5.3.6 Conclusion of structural characterization

Iron oxidic species supported on SBA-15 were analyzed by a combination of various characterization methods. Synthesis procedure using two different Fe(III) precursors led to Fe_xO_y/SBA-15 model catalysts with desired iron loadings. Characteristic two-dimensional hexagonal pore structure of SBA-15 support material remained unaffected by synthesis procedure, as revealed by N_2 physisorption, X-ray diffraction, and TEM measurements. Iron oxidic species were small and highly dispersed in the pore system of SBA-15. A slight expansion of the hexagonal unit cell of SBA-15 after supporting iron oxidic species indicated successful insertion of Fe_xO_y species in the pores of SBA-15. Both formation of crystalline particles and monolayer coverage was excluded based on the absence of diffraction peaks in XRD patterns and the lack of observable Fe_xO_y particles in TEM micrographs. Iron oxide surface coverages were furthermore well below monolayer coverage.

An iron loading-dependent particle size effect was observed for Fe_xO_y/SBA-15 samples independent of the precursor. Increasing iron loading induced an increasing iron oxidic species size. This iron loading-dependent particle size effect was confirmed by DR-UV-Vis, Raman, and Mössbauer spectroscopy. Furthermore, refinement of the broad absorption in DR-UV-Vis spectra Fe_xO_y/SBA-15 samples together with results from Mössbauer spectroscopy suggested a bimodal particle size distribution for Fe_xO_y/SBA-15. Fe_xO_y species were predominantly dimeric or small oligomers with octahedrally coordinated Fe^{3+}. Octahedral coordination sphere of Fe^{3+} was revealed by XANES analysis in addition to DR-UV-Vis and Mössbauer spectroscopy results. Bimodal particle size distribution was ascribed to the presence of Fe_xO_y oligomers with two different sizes. However, small amounts of α-Fe_2O_3-like species and isolated octahedrally coordinated Fe^{3+} species were also observed.

Additionally, structural characterization of Fe_xO_y/SBA-15 samples revealed precursor-dependent differences in surface and porosity characteristics. The citrate precursor induced significantly smaller iron oxidic species, accompanied by a higher degree of micropore filling and a smoother support surface compared to the nitrate precursor. These differences were ascribed to a more pronounced chelating effect of the citrate precursor compared to the nitrate precursor. Results from Raman and Mössbauer spectroscopy, indicating α-Fe_2O_3-like and β-FeOOH-like species being present in Fe_xO_y/SBA-15 samples, were corroborated by detailed X-ray absorption fine structure analysis. The citrate precursor induced a higher amount of β-FeOOH-like species in Fe_xO_y/SBA-15 samples. However, nitrate samples also showed presence of α-Fe_2O_3-like and β-FeOOH-like species, whereas amount of β-FeOOH-like species was significantly lower. The amount of β-FeOOH-like species in Fe_xO_y/SBA-15 samples was strongly dependent on iron loading. Decreasing iron loading correlated with increasing amount of β-FeOOH-like species independent of the precursor. Differences in phase composition of Fe_xO_y/SBA-15 samples could be explained by a chemical memory of supported iron oxidic species on SBA-15. Apparently, chemical environment of the precursor during synthesis hardly affected iron oxidic species after calcination. Using the citrate precursor effected strong interactions between precursor molecules and SBA-15 support material according to the pronounced chelating effect. Hence, hydrogen bonds between silanol groups of SBA-15 and citrate precursor, being located in the pore system of SBA-15, were formed. Conversely, amount of hydrogen bonds of the nitrate precursor was minor due to the less pronounced chelating effect. These differences in amount of hydrogen bonds during synthesis were reflected in phase fractions of resulting Fe_xO_y/SBA-15 samples.

Table 5.8: Overview of results from structural characterization of Fe_xO_y/SBA-15 samples dependent on the precursor. ✓: revealed for both. >: more pronounced/higher for citrate samples. <: more pronounced/higher for nitrate samples.

	Citrate samples	Nitrate samples
Successful synthesis	✓	
Absence of crystalline particles	✓	
Highly dispersed Fe_xO_y	✓	
Iron loading-dependent particle size effect	✓	
Symmetric coordination sphere	✓	
Octahedrally coordinated Fe^{3+}	✓	
Bimodal particle size distribution	✓	
Micropore filling	>	
Chelating effect of Fe(III) precursor	>	
Amount of β-FeOOH-like species	>	
Slight expansion of SBA-15 unit cell	<	
Size of Fe_xO_y species	<	
Smoothing of SBA-15 surface	>	

5.4 Reducibility of Fe_xO_y/SBA-15 catalysts

5.4.1 Importance of reducibility for Fe_xO_y/SBA-15 catalysts

Selective oxidation of propene catalyzed by metal oxide catalysts is assumed to proceed according to the so-called Mars-van-Krevelen mechanism. [63] During this redox-type mechanism, catalysts are partially reduced and re-oxidized. Therefore, reducibility and redox properties of the metal oxide catalysts have a crucial influence on catalytic performance. [7, 27, 30, 146, 147] Understanding reduction and re-oxidation behavior of metal oxide catalysts is a fundamental starting point for deducing reliable structure-activity correlations. Reducibility of Fe_xO_y/SBA-15 catalysts was investigated by temperature-programmed reduction (TPR) experiments with hydrogen as reducing agent. In this chapter, influence of various precursors on reducibility of supported iron oxidic species on SBA-15 will be highlighted. Furthermore, applicability of solid-state kinetic analysis to supported iron oxidic species on high surface support material SBA-15 will be shown.

5.4.2 Temperature-programmed reduction with hydrogen

Figure 5.27 and Figure 5.28 depict TPR traces of Fe_xO_y/SBA-15 catalysts measured during reduction with 5% H_2 in argon at a heating rate of 10 K/min. Significant differences in reduction profiles were discernible. Lowest loaded citrate samples up to an iron loading of 3.0wt% and lowest loaded nitrate sample possessed one single reduction peak. This was indicative of a single-step reduction mechanism. Conversely, higher loaded citrate samples showed a two-step reduction (not considering a very small third TPR peak for sample 10.7wt% Fe_Citrate), while higher loaded nitrate samples showed a three-step reduction at iron loadings higher than 3.3wt%. The first reduction step was assigned to reduction of Fe(III) oxidic species to Fe(II) oxidic species. Apparently, small iron oxidic species of lowest loaded citrate samples and 2.0wt% Fe_Nitrate prevented further reduction in the applied temperature range. Hence, these samples showed only one single reduction peak in the TPR profile. Conversely, the larger iron oxidic species of the higher loaded citrate and nitrate samples exhibited further reduction of Fe(II) oxidic species, and hence, a two-step or even three-step reduction mechanism. Furthermore, nitrate samples showed a shift of the first TPR maxima to lower temperatures compared to citrate samples with similar iron loading. This shift of the TPR maxima indicated better reducibility of the nitrate samples. Accordingly, larger iron oxidic species obtained from nitrate precursor correlated with a better reducibility compared to corresponding citrate samples.

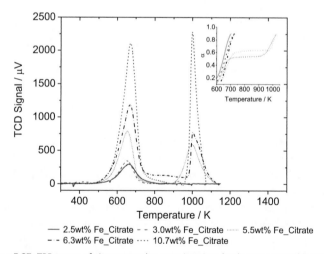

Figure 5.27: TPR traces of citrate samples at various iron loadings measured in 5% H_2 in argon at 10 K/min. Inset depicts reduction degree traces from left to right with increasing iron loading.

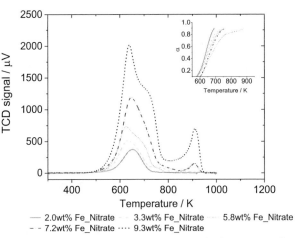

Figure 5.28: TPR traces of nitrate samples at various iron loadings measured in 5% H_2 in argon at 10 K/min. Inset depicts reduction degree traces from left to right with increasing iron loading.

Additionally, reduction degree α traces were extracted by integration of TPR traces for all Fe_xO_y/SBA-15 samples. Resulted α traces are depicted in the inset in Figure 5.27 and Figure 5.28. Lowest loaded citrate samples, with iron loadings from 2.5 through 5.5w%, possessed characteristic sigmoidal α traces as expected for non-isothermal reduction conditions. However, highest loaded citrate samples with iron loadings higher than 5.5wt% showed α traces being divided into two sigmoidal parts, which was indicative of a distinct two-step reduction mechanism. α traces of all nitrate samples showed characteristic sigmoidal behavior independent of the iron loading.

In addition to Fe_xO_y/SBA-15 catalysts, TPR measurements of the mechanical mixture Fe_2O_3/SBA-15 and crystalline α-Fe_2O_3 were conducted at 10 K/min and resulted TPR traces are depicted in Figure 5.29. The mechanical mixture Fe_2O_3/SBA-15 exhibited two TPR peaks with a shoulder at the second peak, indicating a three-step reduction mechanism. TPR traces of the mechanical mixture differed significantly from those of the Fe_xO_y/SBA-15 samples. Moreover, neither the Fe_xO_y/SBA-15 samples nor the mechanical mixture showed a TPR profile characteristic for crystalline α-Fe_2O_3. Differences in TPR profiles of the mechanical mixture and crystalline α-Fe_2O_3 resulted from differences in both particle sizes and dispersion of α-Fe_2O_3 crystallites. [41] Dispersion of smaller α-Fe_2O_3 crystallites on SBA-15 in the mechanical mixture compared to pure α-Fe_2O_3 induced a decreased first TPR peak and a shift of the second TPR peak to lower temperature.

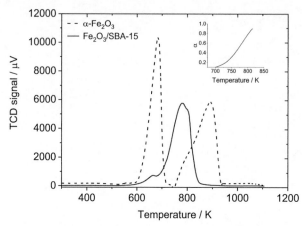

Figure 5.29: TPR traces of mechanical mixture Fe$_2$O$_3$/SBA-15 and crystalline α-Fe$_2$O$_3$ measured in 5% H$_2$ in argon at 10 K/min. Inset depicts reduction degree trace of the mechanical mixture Fe$_2$O$_3$/SBA-15.

5.4.3   Solid-state kinetic analysis of Fe$_x$O$_y$/SBA-15 under non-isothermal conditions

Introduction

Investigations of structure-activity correlations in heterogeneous catalysis are frequently performed using time-resolved measurements under non-isothermal conditions. Hence, additional solid-state kinetic analysis of experimental data measured under these conditions may be helpful in corroborating correlations of structure and function of heterogeneous catalysts. [82–85] In addition to conventional characterization (see chapter 5.3 and 5.4.2), solid-state kinetic analysis was applied to elucidate correlations between precursor, iron loading, and reducibility. In the following, a detailed solid-state kinetic analysis of the reduction traces will be presented. Besides TPR traces of all Fe$_x$O$_y$/SBA-15 samples, those of the mechanical mixture were analyzed. After transforming TPR traces to reduction degree α traces, model-independent and model-dependent solid-state kinetic analysis methods were applied.

Concerning small and highly dispersed Fe$_x$O$_y$ species in the pore system of SBA-15, diffusional effects might control the total reduction process. For this reason, diffusional effects of reactant gas, H$_2$, and product, H$_2$O, had to be excluded to ensure meaningful solid-state kinetic analysis. All Fe$_x$O$_y$/SBA-15 samples showed symmetrically shaped TPR profiles (Figure 5.27-Figure 5.29). This indicated no rate limitation by removal of the small amount of H$_2$O formed by reduction of the low concentration of iron oxidic species on SBA-15. Additionally, mass transport limited processes exhibit characteristic apparent activation energies of less than 10 kJ/mol. [148] Apparent activation energies for all Fe$_x$O$_y$/SBA-15 samples were significantly higher than 10 kJ/mol (Table 5.10). Therefore, mass transport limitation of reactant gas H$_2$ was considered not to be rate-limiting in reduction of Fe$_x$O$_y$/SBA-15.

For solid-state kinetic analysis of reduction of Fe_xO_y/SBA-15 samples and the mechanical mixture under non-isothermal conditions, TPR measurements in 5% H_2 in argon were conducted at varying heating rate between 5 and 20 K/min. TPR traces together with therefrom extracted reduction degree α traces as function of temperature at various heating rates are exemplary depicted for sample 7.2wt% Fe_Nitrate (Figure 5.30). Both TPR traces and reduction degree α traces were shifted to higher temperatures with increasing heating rate. This shift of TPR traces and reduction degree α traces dependent on heating rate was observed for all Fe_xO_y/SBA-15 samples and for the mechanical mixture.

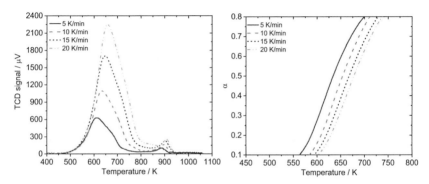

Figure 5.30: TPR traces (left) and reduction degree α traces (right) as function of temperature for reduction (5% H_2 in argon) of 7.2wt% Fe_Nitrate at various heating rates (5, 10, 15, and 20 K/min).

Fundamentals of solid-state kinetic analysis
In general, the rate of a solid-state reaction is given by

$$\frac{d\alpha}{dt} = k(T) \cdot f(\alpha)$$

(5.6)

with reaction degree, α, time, t, temperature-dependent rate constant, $k(T)$, absolute temperature, T, and differential reaction model, $f(\alpha)$. Integration of equation (5.6) yields the integral rate law

$$g(\alpha) = k(T) \cdot t$$

(5.7)

where $g(\alpha)$ is the integral reaction model. Temperature dependence of the rate constant is expressed by the Arrhenius equation:

$$k(T) = A \cdot e^{-\frac{E_a}{RT}}$$

(5.8)

with preexponential (frequency) factor, A, apparent activation energy of rate-determining step, E_a, and gas constant, R. Considering the temperature dependence of the rate constant by substituting equation (5.8) in equation (5.6) and (5.7) yields

$$\frac{d\alpha}{dt} = A \cdot e^{-\frac{E_a}{RT}} \cdot f(\alpha) \tag{5.9}$$

and

$$g(\alpha) = A \cdot e^{-\frac{E_a}{RT}} \cdot t . \tag{5.10}$$

Kinetic parameters, i.e. the "kinetic triple" (apparent activation energy, E_a, preexponential (frequency)factor, A, and suitable solid-state reaction model, $g(\alpha)$), can be obtained by analysis of isothermal data, based on rate laws expressed by equations (5.9) and (5.10). [82–84, 149] Conversely, for analysis of kinetic data measured under non-isothermal reaction conditions, heating rate, β, has to be considered. For non-isothermal measurements a constant heating rate is generally applied. For introducing the heating rate, isothermal reaction rate has to be transformed to non-isothermal reaction rate, by expanding with dT/dT:

$$\frac{d\alpha}{dT} := \frac{d\alpha}{dt} \frac{dt}{dT} \tag{5.11}$$

with non-isothermal reaction rate, $d\alpha/dT$, isothermal reaction rate, $d\alpha/dt$, and reciprocal heating rate, dt/dT. Differential expression of the non-isothermal rate law results from substitution of equation (5.9) into equation (5.11):

$$\frac{d\alpha}{dT} = \frac{A}{\beta} \cdot e^{-\frac{E_a}{RT}} \cdot f(\alpha). \tag{5.12}$$

The differential expression of the non-isothermal rate law can be transferred into the integral expression by separating variables and integrating

$$g(\alpha) = \frac{A}{\beta} \int_0^T e^{-\frac{E_a}{RT}} dT . \tag{5.13}$$

Equation (5.13) is denoted temperature integral and cannot be solved analytically. [82, 150] For transferring the temperature integral into a general form, integration variable x is defined as

$$x = \frac{E_a}{RT} . \tag{5.14}$$

After substituting x in equation (5.13) and changing the integration boundaries, the temperature integral is given as

$$g(\alpha) = \frac{AE_a}{\beta R} \int_x^\infty \frac{e^{-x}}{x^2} dx. \tag{5.15}$$

Furthermore, applying the exponential integral $p(x)$

$$p(x) = \int_x^\infty \frac{e^{-x}}{x^2} \, dx \tag{5.16}$$

yields the temperature integral in the general expression

$$g(\alpha) = \frac{AE_a}{\beta R} \cdot p(x). \tag{5.17}$$

This general expression of the temperature integral can be approximated by different approaches. In addition to numerical calculations of the exponential integral, $p(x)$ can be transformed into an approximated expression that can be integrated. Furthermore, series expansions can be applied for approximating $p(x)$. [82, 89, 150] Asymptotic series expansion and Schlömilch series expansion are the most frequently applied approaches for approximating the exponential integral. Asymptotic series expansion approximates $p(x)$ by partial integration:

$$p(x) = \frac{e^{-x}}{x^2} \left[1 - \left(\frac{2!}{x}\right) + \left(\frac{3!}{x^2}\right) - \left(\frac{4!}{x^3}\right) + \cdots + (-1)^n \left(\frac{(n+1)!}{x^n}\right) + \cdots \right]. \tag{5.18}$$

According to Schlömilch series expansion, exponential integral is approximated by

$$p(x) = \frac{e^{-x}}{x(x+1)} \left[1 - \left(\frac{1}{(x+2)}\right) + \left(\frac{2}{(x+2)(x+3)}\right) \right.$$
$$\left. - \left(\frac{4}{(x+2)(x+3)(x+4)}\right) + \cdots \right]. \tag{5.19}$$

Commonly, Doyle and Senum-Yang approximations are used for approximating the temperature integral. For Doyle approximation, the first three terms of the Schlömilch series expansion (eq. (5.19)) are applied for x values of $20 < x < 60$. A nearly linear relationship between $log(p(x))$ and x over the considered x range was observed by Doyle. [82, 89, 151] Doyle approximation of the temperature integral is given by

$$log \, p(x) \approx -2.315 - 0.457x. \tag{5.20}$$

However, Doyle approximation for x values outside the range $20 < x < 60$ is not valid. Senum-Yang approximation constitutes an alternative for Doyle approximation. Accordingly, the temperature integral is non-linearly approximated based on rational approximation of Luke. [152] The coefficients for different polynomial degrees, n, developed by Luke, are applied in Senum-Yang approximation of the temperature integral. Table 5.9 summarizes Senum-Yang approximations of the temperature integral for various degrees. [153, 154]

Table 5.9: Senum-Yang approximations of the temperature integral for various degrees.

Degree	$p(x)$
1	$\dfrac{e^{-x}}{x}\dfrac{1}{(x+2)}$
2	$\dfrac{e^{-x}}{x}\dfrac{(x+4)}{(x^2+6x+6)}$
3	$\dfrac{e^{-x}}{x}\dfrac{(x^2+10x+18)}{(x^3+12x^2+36x+24)}$
4	$\dfrac{e^{-x}}{x}\dfrac{(x^3+18x^2+86x+96)}{(x^4+20x^3+120x^2+240x+120)}$

Kissinger method

Kissinger method is a model-independent method for analyzing non-isothermal kinetic data. This method is based on the second derivative of the differential expression of the non-isothermal rate law (eq. (5.12)). The second derivative becomes zero exactly when the reaction rate is maximal. Mathematical expression of the Kissinger method is given by

$$ln\frac{\beta}{T_m^2} = ln\left(\frac{AR(n(1-\alpha)_m^{n-1})}{E_a}\right) - \frac{E_a}{RT_m} \tag{5.21}$$

with temperature at maximal reaction rate, T_m, heating rate, β, pre-exponential (frequency) factor, A, gas constant, R, reaction degree at maximal reaction rate, α_m, and apparent activation energy of rate-determining step, E_a. Kissinger method assumes a constant apparent activation energy for the entire reaction degree. Therefore, Kissinger method cannot yield a detailed evolution of apparent activation energy for rate-determining step in a multi-step reaction mechanism. However, Kissinger method does not require any model assumptions and is therefore standardly applied for solid-state kinetic analysis in order to determine a constant value of apparent activation energy for the entire investigated reaction degree. [82, 87]

Apparent activation energy E_a of rate-determining step during reduction of Fe$_x$O$_y$/SBA-15 samples was determined by applying the Kissinger method. Therefore, $ln[\beta/T_m^2]$ was depicted as function of $1000/T_m$ as exemplary shown for sample 7.2wt% Fe_Nitrate (Figure 5.31). [82, 87] Here, T_m corresponded to the first maximum of the TPR traces (Figure 5.27 - Figure 5.29). From the slope of the resulting straight line, the apparent activation energy for the reduction of Fe$_x$O$_y$/SBA-15 was calculated (Figure 5.31). Table 5.10 summarizes apparent activation energies for Fe$_x$O$_y$/SBA-15 samples and the mechanical mixture.

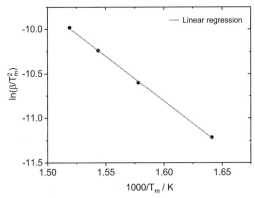

Figure 5.31: Kissinger plot for sample 7.2wt% Fe_Nitrate extracted from TPR traces measured during reduction in 5% H_2 in argon.

Citrate and nitrate samples showed an opposite trend in apparent activation energy as function of iron loading (Figure 5.32). For the nitrate samples, highest apparent activation energy of approximately 90 kJ/mol within the experimental errors was calculated for lowest loaded samples with an iron loading from 2.0 through 5.8wt%. Increasing the iron loading of the nitrate samples resulted in a decreasing apparent activation energy of rate-determining step in reduction. Moreover, results of Kissinger method correlated with the iron loading-dependent particle size effect revealed from DR-UV-Vis and Mössbauer spectroscopy (see chapter 5.3.3). Hence, increasing size of iron oxidic species with increasing iron loading within the nitrate samples was accompanied by better reducibility and decreasing apparent activation energy of reduction. The apparent activation energy of the mechanical mixture was calculated to be 59 ± 7 kJ/mol. This lower apparent activation energy compared to the nitrate samples was consistent with a further increased species size (Table 5.10). Accordingly, larger iron oxidic species, possessing less interactions with SBA-15 support material, showed a better reducibility. This correlation agreed with the reported better reducibility of silica supported iron oxide catalysts with increasing degree of aggregation of Fe_xO_y oligomers. [48, 50, 56] Conversely, for the citrate samples, lowest loaded citrate sample possessed lowest apparent activation energy of 39 ± 8 kJ/mol. Increasing iron loading within the citrate samples resulted in an increasing apparent activation energy of rate-determining step in reduction. This opposite trend of citrate and nitrate samples in apparent activation energy as function of iron loading indicated significant differences in iron oxidic species dependent on the precursor (see chapter 5.3.4). The citrate precursor induced higher amounts of β-FeOOH-like species and the amount of β-FeOOH-like species increased with decreasing iron loading. Conversely, the nitrate precursor yielded higher amounts of α-Fe_2O_3-like species. These precursor-dependent differences in types of iron oxidic species might be responsible for the different trend of apparent activation energy in reduction as function of iron loading. α-Fe_2O_3-like and β-FeOOH-like species possessed differences in local structure. Bond length of Fe-O bonds were shorter than those of Fe-OH bonds (see chapter 5.3.4). Furthermore, Fe-OH bonds in the β-FeOOH crystal structure are described as weak bonds. [139] The presence of weaker Fe-OH bonds in β-FeOOH-like species might be correlated with a better reducibility of iron oxidic species. Hence, higher amounts of β-FeOOH-like species for lower loaded citrate samples might have induced a better reducibility, and hence, a lower apparent activation energy of rate-determining step in reduction compared to higher loaded citrate samples. Especially at low iron loading, the amount of β-FeOOH-like species seemed to be crucial for reducibility, while for higher iron loadings, the influence of interactions between iron oxidic species and support material was predominant. For higher loaded citrate samples, a more pronounced micropore filling accompanied by a smoother SBA-15 surface was observed by N_2 physisorption measurements (see chapter 5.3.1). This was indicative of stronger interactions between iron oxidic species and SBA-15 support material for higher iron loadings, and accordingly, complied with an increased apparent activation energy in reduction.

Furthermore, precursor-dependent differences in values of apparent activation energy were observed. The citrate samples possessed higher values of apparent activation energy, except for 2.5wt% Fe_Citrate, compared to those of nitrate samples. This complied with observed precursor-dependent particle size effect (see chapter 5.3.3). Smaller Fe_xO_y species obtained from citrate precursor induced a higher degree of micropore filling accompanied by a smoother support surface. Hence, small and highly dispersed iron oxidic species possessed strong interactions with SBA-15 support material. Consequently, reducibility was hindered and apparent activation energy of rate-determining step in reduction was higher for citrate samples compared to nitrate samples with similar iron loading (Figure 5.32).

Table 5.10: Apparent activation energy, E_a, of rate-determining step in reduction of Fe_xO_y/SBA-15 catalysts and mechanical mixture Fe_2O_3/SBA-15. E_a values were determined by Kissinger method and reduction was conducted in 5% H_2 in argon.

	E_a / kJ/mol
2.5wt% Fe_Citrate	39 ± 8
3.0wt% Fe_Citrate	96 ± 3
5.5wt% Fe_Citrate	94 ± 1
6.3wt% Fe_Citrate	113 ± 8
10.7wt% Fe_Citrate	104 ± 8
2.0wt% Fe_Nitrate	88 ± 8
3.3wt% Fe_Nitrate	90 ± 4
5.8wt% Fe_Nitrate	90 ± 2
7.2wt% Fe_Nitrate	84 ± 1
9.3wt% Fe_Nitrate	62 ± 8
Fe_2O_3/SBA-15	59 ± 7

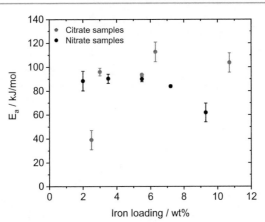

Figure 5.32: Apparent activation energy, E_a, of rate-determining step in reduction (5% H_2 in argon) determined by Kissinger method for Fe_xO_y/SBA-15 catalysts at various iron loadings.

Method of Ozawa, Flynn, and Wall

A single apparent activation energy value resulting from Kissinger method may not be sufficient for a detailed kinetic analysis of a solid-state reaction. Therefore, the isoconversional, model-independent method of Ozawa, Flynn, and Wall (OFW) was applied for determining the evolution of apparent activation energy of rate-determining step as function of reduction degree α. [82, 88, 89, 150] The OFW method uses the decade logarithm of the non-isothermal temperature integral (equation (5.17)) and the Doyle approximation (equation (5.20)) for $log(p(x))$ resulting in

$$\log(g(\alpha)) = log\left(\frac{AE_a}{\beta R}\right) - 2.315 - 0.457x \qquad (5.22)$$

with integral solid-state reaction model, $g(\alpha)$, pre-exponential (frequency) factor, A, apparent activation energy, E_a, heating rate, β, and gas constant, R. Substituting x according to equation (5.14) and rearranging yields

$$\log(\beta) = log\left(\frac{A_\alpha E_{a,\alpha}}{g(\alpha)R}\right) - 2.315 - 0.457\frac{E_{a,\alpha}}{R\,T_{\alpha,\beta}} \qquad (5.23)$$

with pre-exponential (frequency) factor at reduction degrees α, A_α, apparent activation energy at reduction degrees α, $E_{a,\alpha}$, and temperatures, $T_{\alpha,\beta}$.

For Fe_xO_y/SBA-15 samples and mechanical mixture Fe_2O_3/SBA-15, the evolution of apparent activation energy, $E_{a,\alpha}$, as function of reduction degree α was determined. Therefore, reduction degree α traces were extracted by integration of the TPR traces measured at various heating rates. First, temperatures $T_{\alpha,\beta}$ for defined reduction degrees α were determined from the experimental α traces at various heating rates. Temperatures $T_{\alpha,\beta}$ were determined for reduction degrees α in the range of 0.1 and 0.8, with $\Delta\alpha = 0.1$. Second, decade logarithm of the heating rate as function of $1000/T_{\alpha,\beta}$ for the different reduction degrees was calculated based on equation (5.23). Figure 5.33 shows the resulting straight lines for heating rates of 5, 10, 15, and 20 K/min and various reduction degrees α for sample 7.2wt% Fe_Nitrate.

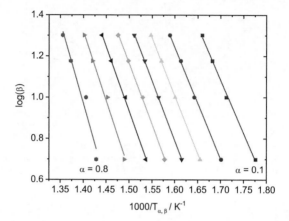

Figure 5.33: Logarithmic heating rate, β, as function of reciprocal temperature for reduction of 7.2wt% Fe_Nitrate in 5% H_2 in argon and reduction degree range between 0.1 and 0.8 (OFW method).

Linear regression of the straight lines resulted in apparent activation energy as function of reduction degree α. Because of $\frac{E_{a,\alpha}}{R\,T_{\alpha,\beta}} < 20$, the apparent activation energy was corrected according to Senum-Yang. [82, 150, 153] Apparent activation energy as function of reduction degree α together with apparent activation energy determined by Kissinger method for Fe_xO_y/SBA-15 samples is depicted in Figure 5.34 for the citrate samples and in Figure 5.35 for the nitrate samples.

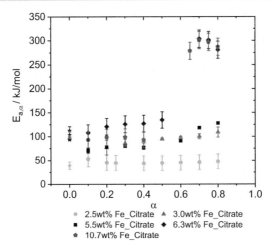

Figure 5.34: Apparent activation energy, $E_{a,\alpha}$, as function of reduction degree, α, for reduction of citrate samples in 5% H_2 in argon according to Senum-Yang approximation. Apparent activation energies determined by Kissinger method are depicted at $\alpha = 0$.

Apparent activation energy obtained from Kissinger method for the citrate samples agreed with apparent activation energy obtained from OFW method (Figure 5.34). Furthermore, the trend in apparent activation energy as function of iron loading revealed by Kissinger method was confirmed by OFW method in α range from 0.1 through 0.5. Accordingly, increasing iron loading within the citrate samples correlated with increasing apparent activation energy in reduction. Apparent activation energies $E_{a,\alpha}$ of lowest loaded citrate samples, 2.5wt% Fe_Citrate and 3.0wt% Fe_Citrate, were invariant in the α range within the error limits. Thus, a single-step reduction mechanism was assumed for these Fe_xO_y/SBA-15 samples coinciding with the single reduction peak in corresponding TPR profiles (Figure 5.27). Such a reaction mechanism is more similar to homogeneous kinetics than to complex heterogeneous kinetics. Conversely, for iron loadings higher than 3.0wt%, significant differences in evolution of apparent activation energy of citrate samples were discernible. The significant increase in $E_{a,\alpha}$ from a reduction degree of 0.6 for sample 5.5wt% Fe_Citrate indicated a change in rate-determining step during a more complex reduction mechanism. [155, 156] Furthermore, evolution of apparent activation energy of 6.3wt% Fe_Citrate and 10.7wt% Fe_Citrate showed a division into two domains. The first domain of $E_{a,\alpha}$ ranged from reduction degree α of 0.1 through 0.5, whereas the second domain ranged from α of 0.6 through 0.8. Values of $E_{a,\alpha}$ were significantly increased for the second domain. The evolution of apparent activation energy as function of reduction degree was indicative of a two-step reduction mechanism comprising a change in rate-determining step during reduction. Such a two-step reduction mechanism for citrate samples with iron loadings higher than 3.0wt% coincided with the pronounced two reduction peaks in the TPR profiles of these samples (Figure 5.27). However, evolution of apparent activation energy as function of reduction degree within the first domain, α range from 0.1 through 0.5, was invariant for 5.5wt% Citrate, 6.3wt% Fe_Citrate, and 10.7wt% Fe_Citrate. For reduction degrees up to 0.5, this invariant $E_{a,\alpha}$ was indicative of a first reduction step with high similarities to homogeneous kinetics. In homogeneous kinetics, isolated ions in solution interact weakly with each other. Accordingly, small iron oxidic species in the pore system of SBA-15 reacted similar to isolated ions in homogeneous solutions.

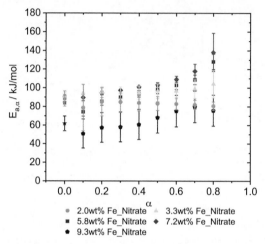

Figure 5.35: Apparent activation energy, $E_{a,\alpha}$, as function of reduction degree, α, for reduction of nitrate samples in 5% H_2 in argon according to Senum-Yang approximation. Apparent activation energies determined by Kissinger method are depicted at $\alpha = 0$.

For the nitrate samples, the apparent activation energy obtained from Kissinger method also agreed with the apparent activation energy obtained from OFW method (Figure 5.35). As already revealed by Kissinger method, highest loaded nitrate sample, 9.3wt% Fe_Nitrate, possessed lowest apparent activation energy $E_{a,\alpha}$ in reduction. Dependent on iron loading, significant differences in evolution of apparent activation energy as function of reduction degree α were observed for the nitrate samples. Lowest loaded nitrate samples, 2.0wt% Fe_Nitrate and 3.3wt% Fe_Nitrate, showed an invariant evolution of apparent activation energy $E_{a,\alpha}$ indicative of a single-step reduction mechanism. This was in accordance with one single reduction peak in TPR profiles of these nitrate samples (Figure 5.28). Compared to lowest loaded samples, for higher loaded nitrate samples with iron loadings higher than 3.3wt%, evolution of apparent activation energy as function of reduction degree differed significantly. The increase of apparent activation energy indicated a change in rate-determining step during a more complex reduction mechanism. [155] Such a more complex reduction mechanism coincided with the multi-step TPR profile for samples 5.8wt% Fe_Nitrate, 7.2wt% Fe_Nitrate, and 9.3wt% Fe_Nitrate (Figure 5.28). Apparently, larger iron oxidic species of higher loaded nitrate samples induced a multi-step reduction mechanism.

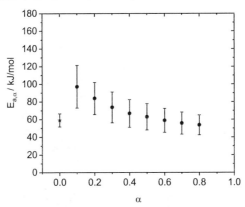

Figure 5.36: Apparent activation energy, $E_{a,\alpha}$, as function of reduction degree, α, for reduction of the mechanical mixture in 5% H_2 in argon according to Senum-Yang approximation. Apparent activation energy determined by Kissinger method is depicted at $\alpha = 0$ (star).

In addition to Fe_xO_y/SBA-15 samples, evolution of apparent activation energy as function of reduction degree was calculated for the mechanical mixture Fe_2O_3/SBA-15 (Figure 5.36). Evolution of apparent activation energy for the mechanical mixture differed significantly from that of Fe_xO_y/SBA-15 samples. At low reduction degrees, $E_{a,\alpha}$ slightly decreased with increasing α. However, for reduction degrees higher than 0.3, evolution of apparent activation energy as function of α was invariant and agreed with the apparent activation energy obtained from Kissinger method. The initial decrease in $E_{a,\alpha}$ was ascribed to the complex reduction mechanism and the less resolved two reduction peaks in the TPR profile of Fe_2O_3/SBA-15 (Figure 5.29). Deviations in $E_{a,\alpha}$ of the mechanical mixture from those of the Fe_xO_y/SBA-15 samples seemed reasonable due to the presence of α-Fe_2O_3 crystallites in the mechanical mixture compared to small, highly dispersed Fe_xO_y species in the Fe_xO_y/SBA-15 samples.

Coats-Redfern method
In addition to model-independent Kissinger and OFW methods, model-dependent Coats-Redfern method [157] provided a complementary analysis of non-isothermal kinetic data. Compared to a model-independent kinetic analysis, model-dependent analysis enables a more detailed characterization of the reaction mechanism. Coats-Redfern method is based on the integral form of non-isothermal rate law (eq. (5.17)) with the temperature integral being approximated by asymptotic series expansion. Coats-Redfern method can be expressed by

$$ln\left(\frac{g(\alpha)}{T^2}\right) = ln\left(\frac{AR}{\beta E_a}\left[1 - \left(\frac{2RT}{E_a}\right)\right]\right) - \frac{E_a}{RT}, \qquad (5.24)$$

with integral solid-state reaction model, $g(\alpha)$, temperature, T, heating rate, β, apparent activation energy of rate-determining step, E_a, gas constant, R, and pre-exponential (frequency) factor, *A*. Accordingly, resulting apparent activation energies are based on assuming a suitable solid-state reaction model. Plotting $ln[g(\alpha)/T^2]$ as function of reciprocal temperature results in straight lines for suitable solid-state reaction models. Apparent activation energies can be obtained from the slope of resulting straight lines. [149, 157]

Linear regression was conducted to determine the apparent activation energy of Fe_xO_y/SBA-15 catalysts. Here, only solid-state reaction models, $g(\alpha)$, resulting in both apparent activation energies similar to those obtained from model-independent methods and good linear regressions were selected for further analysis. For investigating the reduction of Fe_xO_y/SBA-15 catalysts and the mechanical mixture Fe_2O_3/SBA-15, reduction degree α curves were analyzed. Applied solid-state reaction models were nucleation models, including power law models (P) and Avrami-Erofeyev models (A), as well as the autocatalytic Prout-Tompkins model (B1). Furthermore, diffusion models (D), geometrical contraction models (R), and reaction order-based models (F) were tested. [149] In the following, analysis procedure according to Coats-Redfern method is exemplary described for lowest loaded citrate sample, consisting of smallest iron oxidic species. These smallest iron oxidic species might cause difficulties by applying solid-state reaction models which are conventionally applied to bulk systems. Nevertheless, analysis procedure according to Coats-Redfern method was successfully applied for lowest loaded citrate sample. Analysis procedure for all other Fe_xO_y/SBA-15 samples was conducted analogously. Only D4, F1, A2, R2, and B1 solid-state reaction models revealed wide linear ranges by plotting $ln[g(\alpha)/T^2]$ as function of reciprocal temperature for sample 2.5wt% Fe_Citrate. Linear ranges of reduction degree α curves using the F1 solid-state reaction model for 2.5wt% Fe_Citrate are depicted in Figure 5.37. Apparent activation energies for solid-state reaction models D4, F1, A2, R2, and B1 as obtained from the slope of resulting straight lines are given in Table 5.11.

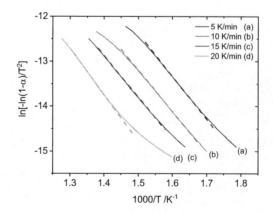

Figure 5.37: Linear ranges of reduction degree α curves using the F1 solid-state reaction model for sample 2.5wt% Fe_Citrate (straight lines). Reduction was conducted in 5% H_2 in argon. Dashed lines: Linear regressions.

Table 5.11: Apparent activation energy of reduction of 2.5wt% Fe_Citrate in 5% H_2 in argon at various heating rates depending on the applied solid-state reaction model.

Heating rate /	E_a / kJ mol^{-1}				
K min^{-1}	D4	A2	B1	R2	F1
5	138.9 ± 0.1	32.6 ± 0.1	43.8 ± 0.2	65.7 ± 0.1	75.0 ± 0.1
10	140.4 ± 0.3	31.9 ± 0.1	41.6 ± 0.3	65.7 ± 0.1	73.7 ± 0.1
15	141.7 ± 0.2	33.6 ± 0.2	45.7 ± 0.1	66.0 ± 0.1	75.7 ± 0.1
20	142.1 ± 0.2	35.7 ± 0.1	54.6 ± 0.3	65.5 ± 0.1	77.9 ± 0.3

Compared to the results of Kissinger and OFW methods, apparent activation energies at different heating rates were significantly higher for the D4 model and significantly lower for the A2 model. Hence, D4 and A2 solid-state reaction models were not considered for further analysis. The B1 model yielded apparent activation energies similar to those obtained from Kissinger and OFW methods. However, the autocatalytic B1 model assumes that defects formed at the reaction interface during nuclei growth further catalyze, and hence, accelerate the reaction. This concept appeared hardly applicable to Fe_xO_y/SBA-15 samples with dispersed iron oxidic species located in a nanostructured pore system. Therefore, the B1 model was not further considered. Similar constraints hold for the R2 model. The R2 reaction model is described as geometrical contraction model in which nucleation occurs on the surface of the cylindrical crystal. Thus, the reaction rate is determined by the decreasing interface area between reactant and product phase during reaction. [149] Again, such a concept seemed not applicable for small and dispersed iron oxidic species on the surface of a porous support. Consequently, the F1 model was chosen as suitable reaction model for 2.5wt% Fe_Citrate and all other Fe_xO_y/SBA-15 samples.

Order-based reaction models are derived from

$$\frac{d\alpha}{dt} = k(1 - \alpha)^n, \tag{5.25}$$

with rate of reaction, $d\alpha/dt$, rate constant, k, reduction degree, α, and reaction order, n. For first-order reaction model, equation (5.25) results in

$$\frac{d\alpha}{dt} = k(1 - \alpha). \tag{5.26}$$

The integral expression for first-order reaction model can be obtained after separating variables and integrating and is expressed by

$$g(\alpha) = -ln(1 - \alpha). \tag{5.27}$$

First-order reaction model, F1, is also known as Mampel reaction model and describes solid-state reactions with a large number of nucleation sites resulting in fast nucleation. Apparently, reduction of Fe_xO_y/SBA-15 samples was neither inhibited by limited mobility of reactants nor by increasing product layer. Order-based reaction models are the simplest solid-state reaction models similar to those used in homogeneous kinetics where ions in solution interact weakly with each other. In homogeneous kinetics, the reaction rate of order-based reaction models is proportional to the concentration of active sites. [149, 158] Because the Fe(III) species of the Fe_xO_y/SBA-15 samples constituted small and isolated nucleation sites, the F1 model can be readily applied to these samples. Moreover, Mampel reaction model constitutes a special case of Avrami-Erofeyev (A) nucleation solid-state reaction models. [149, 158] For $n = 1$ in Avrami-Erofeyev models, the Mampel reaction model considers a large number of nucleation sites with nuclei growing predominantly in one dimension. Despite Fe_xO_y/SBA-15 samples consisting of small and highly dispersed iron oxidic species located in the pore system of SBA-15, application of Mampel (F1) solid-state reaction model was successful and resulted in meaningful results of reduction mechanism.

For the mechanical mixture Fe_2O_3/SBA-15, plotting $ln[g(\alpha)/T^2]$ as function of reciprocal temperature resulted in wide linear ranges by applying reaction model R3 (Figure 5.38). Hence, the R3 solid-state reaction model was a suitable reaction model. The R3 reaction model is denoted as contracting sphere model with nucleation occurring rapidly on the surface of the particles. This reaction model was consistent with a mixture of α-Fe_2O_3 crystallites and SBA-15 support material as obtained by conventional sample characterization (see chapter 5.3). Geometrical contraction models (R) assume a reaction interface progress towards the center of the crystal particle. For any crystal particle, radius of the particle, r, at time t is expressed by

$$r = r_0 - kt \tag{5.28}$$

where r_0 is the radius at time t_0. Dependent on particle shape, different mathematical descriptions are required. For particles with either a spherical or a cubical shape, contracting sphere model R3 is valid. The integral expression for R3 reaction model is given by [149, 159]

$$g(\alpha) = 1 - (1 - \alpha)^{1/3}. \tag{5.29}$$

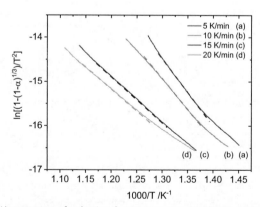

Figure 5.38: Linear ranges of reduction degree α curves using the R3 solid-state reaction model for mechanical mixture Fe_2O_3/SBA-15 (straight lines). Reduction was conducted in 5% H_2 in argon. Dashed lines: Linear regressions.

JMAK kinetic analysis

Conventional model-dependent solid-state kinetic analysis yields a suitable solid-state kinetic reaction model for the investigated reaction. Johnson-Mehl-Avrami-Kolmogorov (JMAK) kinetic analysis additionally enables deducing geometrical information, i.e. topological dimension and Avrami exponent, of solid-state reaction. Therefore, JMAK kinetic analysis was applied to geometrically describe the reduction of Fe_xO_y/SBA-15 samples under non-isothermal conditions. [160–162] JMAK kinetic analysis is based on

$$ln[-\ln(1-\alpha)] = -n\,ln(\beta) - 1.052\frac{mE}{RT} + const. \tag{5.30}$$

with reduction degree, α, Avrami exponent, n, heating rate, β, topological dimension, m, apparent activation energy of rate-determining step, E, gas constant, R, and temperature, T. Plotting *ln[-ln(1-α)]* as function of reciprocal temperature at different heating rates resulted in straight lines, as exemplary depicted for sample 7.2wt% Fe_Nitrate (Figure 5.39, left). From the slope of the resulting straight lines, the topological dimension, m, can be determined. Here, the apparent activation energy obtained from Kissinger method was inserted in equation (5.30). Based on equation (5.30), the Avrami exponent, n, is derived according to

$$-n = \frac{d\{ln[-\ln(1-\alpha)]\}}{d[\ln(\beta)]}\Bigg|_T. \tag{5.31}$$

Thus, values of *ln[-ln(1-α)]* were calculated at fixed temperatures and plotted as function of *ln(β)*. The slopes of the resulting straight lines (Figure 5.39, right) were used to determine the Avrami exponents.

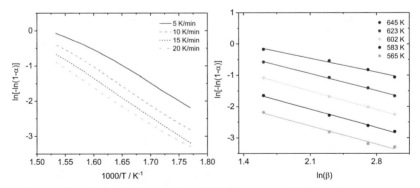

Figure 5.39: Left: *ln[-ln(1-α)]* as function of *1000/T* according to JMAK kinetic analysis for determining topological dimension of reduction of 7.2wt% Fe_Nitrate. Right: *ln[-ln(1-α)]* as function of *ln(β)* according to JMAK kinetic analysis for determining Avrami exponent for 7.2wt% Fe_Nitrate. Reduction was conducted in 5% H_2 in argon.

Plotting *ln[-ln(1-α)]* as function of reciprocal temperature did not afford straight lines for sample 2.5wt% Fe_Citrate. Therefore, JMAK kinetic analysis was not applied to the data of this sample. Topological dimension and Avrami exponent as function of temperature and heating rate for all other Fe_xO_y/SBA-15 samples and Fe_2O_3/SBA-15 are depicted in Figure 5.40 - Figure 5.42.

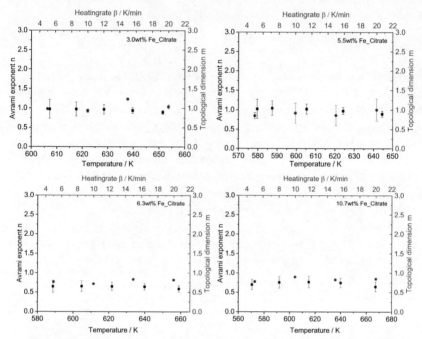

Figure 5.40: Topological dimension (blue circles, right axis) and Avrami exponent (black squares, left axis) from JMAK kinetic analysis as function of temperature and heating rate for citrate samples.

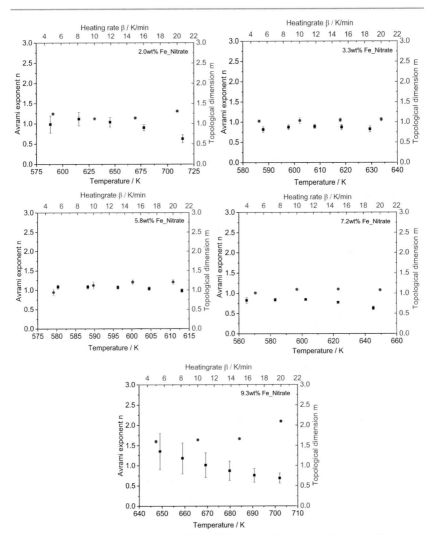

Figure 5.41: Topological dimension (blue circles, right axis) and Avrami exponent (black squares, left axis) from JMAK kinetic analysis as function of temperature and heating rate for nitrate samples.

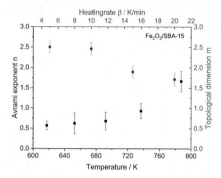

Figure 5.42: Topological dimension (blue circles, right axis) and Avrami exponent (black squares, left axis) from JMAK kinetic analysis as function of temperature and heating rate for the mechanical mixture Fe$_2$O$_3$/SBA-15.

Topological dimension and Avrami exponent for all citrate samples were approximately one, independent of the iron loading (Figure 5.40). One-dimensionality ($m = 1$) of reduction was consistent with small Fe$_x$O$_y$ species being in the pore system of SBA-15. At $n = m = 1$, the reduction mechanism is governed by site saturation. Thus, at the beginning of the reduction nucleation sites either already existed or were formed immediately. [162, 163] Avrami exponent of approximately one and topological dimension of one resulted for nitrate samples with an iron loading from 2.0 through 7.2wt% (Figure 5.41). Accordingly, reduction of nitrate samples with an iron loading up to 7.2wt% was one-dimensional and governed by site saturation, as already observed for the citrate samples. The Coats-Redfern method identified first-order, Mampel, solid-state reaction model being suitable to describe the kinetic data of Fe$_x$O$_y$/SBA-15 samples. Mampel solid-state reaction model was consistent with the reduction being governed by site saturation ($n = m = 1$). Moreover, the Mampel model represents an exception of the Avrami-Erofeyev model with an Avrami exponent of $n = 1$. Hence, results from JMAK kinetic analysis and model-dependent Coats-Redfern method agreed well for all citrate samples and for nitrate samples up to an iron loading of 7.2wt%.

Conversely, highest loaded nitrate sample, 9.3wt% Fe_Nitrate, exhibited an Avrami exponent of approximately one, while topological dimension ranged from 1.5 through 2 (Figure 5.41, bottom). This indicated an increased dimensionality of the reduction for highest loaded nitrate sample. The increase in topological dimension correlated with both larger Fe$_x$O$_y$ species for this sample compared to all other samples and higher amount of α-Fe$_2$O$_3$-like species at this iron loading (see chapter 5.3.3 and 5.3.4). Accordingly, larger Fe$_x$O$_y$ species induced not only a higher contribution of micropores to the entire SBA-15 surface area accompanied by an increased surface roughness, but also an increased dimensionality of reduction reaction. However, topological dimension of 9.3wt% Fe_Nitrate was lower compared to that of the mechanical mixture. This agreed with smaller iron oxidic species and the absence of XRD peaks for 9.3wt% Fe_Nitrate, while the mechanical mixture showed characteristic diffraction peaks of α-Fe$_2$O$_3$ (see chapter 5.3.2). The mechanical mixture Fe$_2$O$_3$/SBA-15 exhibited the highest values of topological dimension. Topological dimension as function of heating rate ranged from 2 through 3 and Avrami exponent increased from 0.5 up to 1.8. These increased values of Avrami exponent and topological dimension were correlated with the presence of α-Fe$_2$O$_3$ crystallites in this sample. The mechanical mixture consisted of α-Fe$_2$O$_3$ crystallites mixed with the support material. Model-dependent Coats-Redfern method identified the geometrical contraction model R3 being a suitable reaction model. Therefore, three-dimensional reduction was compatible with a rapid nucleation on the α-Fe$_2$O$_3$ crystallites. Thus, for the mechanical mixture Fe$_2$O$_3$/SBA-15, results from model-dependent Coats-Redfern analysis were confirmed by JMAK kinetic analysis.

5.4.4 Correlation between solid-state kinetic analysis and structural characterization

Results from structural characterization agreed well with those from solid-state kinetic analysis for all Fe$_x$O$_y$/SBA-15 samples. Smaller iron oxidic species resulting from citrate precursor coincided with a higher degree of micropore filling and a smoother support surface. This indicated stronger interactions between iron oxidic species and SBA-15 (see chapter 5.3.1). Consequently, citrate samples possessed minor reducibility, i.e. higher values of apparent activation energy in reduction compared to nitrate samples for iron loadings higher than 2.5wt%. Larger iron oxidic species obtained from nitrate precursor possessed weaker interactions to SBA-15 support material and induced both a more complex reduction mechanism and a better reducibility (see chapter 5.4.2 and 5.4.3). Precursor-dependent differences in reducibility were furthermore ascribed to precursor-dependent differences in types of iron oxidic species. Higher amounts of β-FeOOH-like species obtained from citrate precursor (see chapter 5.3.4) might have induced the better reducibility of lowest loaded citrate samples.

Moreover, iron loading-dependent differences in reducibility were observed, whereby citrate and nitrate samples showed an opposite behavior. For nitrate samples, an increasing species size with increasing iron loading (see chapter 5.3.3), correlated with a decreasing apparent activation energy of reduction. Accordingly, lower apparent activation energy in reduction of higher loaded nitrate samples coincided with iron loading-dependent particle size effect. Conversely, within the citrate samples, apparent activation energy in reduction decreased with decreasing iron loading. This indicated that types of iron oxidic species were crucial for reducibility at low iron loadings. For higher iron loadings, predominantly interactions between iron oxidic species and support material seemed to account for the reducibility.

Structural characterization analysis methods identified the Fe(III) species as being isolated in the pore system of SBA-15 and interacting weakly with each other. Even for the higher loaded Fe$_x$O$_y$/SBA-15 samples with more aggregated Fe$_x$O$_y$ oligomers, weakly interacting and well-dispersed Fe(III) species were assumed. With respect to solid-state kinetic analysis, iron species in the pores of SBA-15 react similar to isolated ions in a homogeneous solution. Accordingly, a first-order reaction model, Mampel model, was suited best to describe the similarity of the Fe$_x$O$_y$/SBA-15 samples to homogeneous systems. Additionally, JMAK kinetic analysis was consistent with a one-dimensional reduction of iron oxidic species localized in the pore system of SBA-15. The increased dimensionality of reduction of highest loaded nitrate sample, 9.3wt% Fe_Nitrate, agreed with the largest iron oxidic species and the highest amount of α-Fe$_2$O$_3$-like species for this sample (see chapter 5.3.3 and 5.3.4).

Not only for the Fe$_x$O$_y$/SBA-15 samples but also for the mechanical mixture Fe$_2$O$_3$/SBA-15, results from structural characterization agreed with those from solid-state kinetic analysis. According to JMAK kinetic analysis, the fraction of crystalline α-Fe$_2$O$_3$ in Fe$_2$O$_3$/SBA-15 (see chapter 5.3.2), induced a three-dimensional reduction. Hence, reduction was governed by rapid nucleation on the three-dimensional α-Fe$_2$O$_3$ crystallites. This was confirmed by model-dependent analysis yielding a contracting sphere model, R3, with rapid nucleation occurring on the surface of the α-Fe$_2$O$_3$ crystallites, as suitable model for the rate-determining step in reduction.

Apparently, for both supported systems and mechanical mixture, results of conventional characterization and solid-state kinetic analysis corroborated each other. This showed that solid-state kinetic analysis, i.e. non-isothermal reaction conditions and model-dependent as well as model-independent analysis methods, can be successfully applied to supported systems in addition to conventional bulk materials. Time- and temperature-dependent measurements, such as TPR or TG/DTA, are readily used in characterizing supported systems. However, those techniques yield little to no structural details of the supported species. Hence, solid-state kinetic analysis of the already available data can give additional information without additional experimental effort.

5.5 Characterization under catalytic conditions

5.5.1 Introduction

The previous chapters of this work revealed a significant influence of the precursor on structural properties, iron oxidic species types, and reducibility of Fe$_x$O$_y$/SBA-15 catalysts. Applicability of solid-state kinetic analysis was furthermore shown for supported iron oxidic species on SBA-15. Additionally, catalytic performance of Fe$_x$O$_y$/SBA-15 catalysts in selective oxidation of propene was investigated. *In situ* characterization revealed structural and electronic changes of Fe$_x$O$_y$/SBA-15 catalysts, which directly correlated with their catalytic performance. Moreover, results from solid-state kinetic analysis of supported iron oxidic species were correlated with catalytic performance of Fe$_x$O$_y$/SBA-15 catalysts. Detailed comparison of Fe$_x$O$_y$/SBA-15 catalysts obtained from citrate and nitrate precursor emphasized the influence of the precursor on catalytic performance.

5.5.2 Estimation of gas phase diffusion effects and gas phase transport limitations

For mesoporous support materials, such as SBA-15, evaluation of the influence of pore diffusion on catalytic reaction is indispensable. To ensure the absence of gas phase diffusion effects in selective oxidation of propene, the Weisz-Prater criterion [164] was applied. This criterion is particularly useful in heterogeneous catalysis as all required parameters are observable and easily calculated or measured. The Weisz-Prater criterion provides a dimensionless number, C_{WP}:

$$C_{WP} = L_{Cat}^2 \cdot \frac{n+1}{2} \cdot \frac{r_{eff,propene} \cdot \rho_{cat}}{D_{eff} \cdot c_{eff,propene}}$$

(5.32)

with average particle length, L_{cat}, reaction order, n, effective propene reaction rate, $r_{eff,propene}$, average catalyst density, ρ_{cat}, Knudsen diffusivity, D_{eff}, and effective propene concentration, $c_{eff,propene}$. If $C_{WP} < 1$ diffusion effects are negligible. Conversely, for $C_{WP} > 1$ pore diffusion is the rate-determining step in reaction.

For SBA-15 support material with an assumed average particle length, L_{cat}, of 50 μm and a pore diameter, d_{pore}, of 8 nm, Knudsen diffusivity had to be assumed due to $L_{cat} \gg d_{pore}$. The Knudsen diffusivity was calculated according to:

$$D_{eff} = \frac{\tilde{v} \cdot d_{pore}}{3}$$

(5.33)

with average velocity of propene molecules in gas phase, $\tilde{v}$.
The average velocity of propene molecules in gas phase was determined by:

$$\tilde{v} = \sqrt{\frac{8 \cdot R \cdot T}{\pi \cdot M_{propene}}}$$

(5.34)

with gas constant, R, temperature of catalytic measurements, T, and molar weight of propene, $M_{propene}$. The calculated or measured parameters used for the Weisz-Prater criterion are summarized in Table 5.12. Considering all experimental parameters, the Weisz-Prater criterion was calculated to be 0.02. Hence, $C_{WP} = 0.02 < 1$ was valid, and therefore, gas phase diffusion effects on catalytic reaction were excluded.

Table 5.12: Experimental values of parameters required for calculation of the Weisz-Prater criterion.

Parameter	Value
T	653.15 K
$M_{propene}$	42.08 g mol^{-1}
$\tilde{v}$	573.05 m s^{-1}
D_{eff}	1.53*10^{-6} m^2 s^{-1}
$c_{eff,propene}$	4.46*10^{-3} mol l^{-1}
ρ_{cat}	2 g cm^{-3}
n	1
$r_{eff,propene}$	3.395*10^{-5} mol g^{-1} s^{-1}

Besides gas phase diffusion effects, heat or mass transport limitations may influence catalytic reaction of heterogeneous catalysts. To exclude heat or mass transport limitations, Koros-Nowak test was applied. [165] The Koros-Nowak test is an experimental test to identify gas phase transport limitations in measurements of catalytic rates. According to the Koros-Nowak test, reaction rate in the kinetic regime is directly proportional to the concentration of active material. Hence, plotting the logarithm of the reaction rate against the logarithm of the weight of active material, m_{cat}, for various metal loadings with similar metal dispersion and similar conversion should result in a line with a slope of 1. [165–167] Otherwise, measured catalytic rates might be influenced by heat or mass transport limitations. Here, the Koros-Nowak test was performed for selective oxidation of propene for Fe_xO_y/SBA-15 catalysts with various iron loadings at propene conversions between 5 and 10%. The magnitude of the slope of the resulting straight line was 1.06 (Figure 5.43). Hence, determined catalytic data in selective oxidation of propene were assumed to be independent of heat and mass transport limitations.

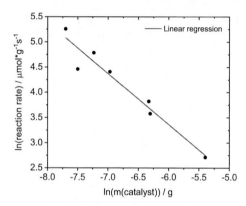

Figure 5.43: Plot for Koros-Nowak test. Logarithm of the reaction rate as function of logarithm of the weight of catalyst for various Fe_xO_y/SBA-15 catalysts.

5.5.3 Catalytic performance in selective oxidation of propene

Catalytic performance of Fe_xO_y/SBA-15 catalysts in selective oxidation of propene was investigated at 653 K and at comparable propene conversions between 5 and 10%. Distribution of main products during selective oxidation of propene for Fe_xO_y/SBA-15 catalysts compared to that of the mechanical mixture are summarized in Table 5.13. Acrolein, carbon oxides, CO and CO_2, and acetaldehyde were the main reaction products. The mechanical mixture, Fe_2O_3/SBA-15, exhibited a very high selectivity towards CO_2. Compared to Fe_xO_y/SBA-15 catalysts, the mechanical mixture showed no significant selectivity towards acrolein and comparatively low reaction rate of propene conversion of 4.8 µmol (g s)$^{-1}$. This low acrolein selectivity of the mechanical mixture suggested crystalline α-Fe_2O_3 particles being not catalytically active. For Fe_xO_y/SBA-15 catalysts, acrolein selectivity as function of time on stream is depicted in Figure 5.44. Dependent on the precursor, significant differences in acrolein selectivity, and furthermore, in catalytic performance were revealed. The citrate samples showed a significantly higher selectivity towards acrolein at iron loadings up to 7.2wt%. Conversely, at higher iron loadings, the nitrate samples showed a higher acrolein selectivity. For citrate samples, highest acrolein selectivity resulted for lowest loaded samples, whereas highest acrolein selectivity for nitrate samples was observed for highest iron loading.

Optimal surface coverage for maximal acrolein selectivity for the citrate samples was 0.47 Fe atoms per nm². For the nitrate samples, optimal surface coverage shifted to a higher value of 1.58 Fe atoms per nm².

Table 5.13: Distribution of main products during selective oxidation of propene for Fe_xO_y/SBA-15 catalysts and the mechanical mixture. Catalytic measurements were conducted in 5% propene, 5% oxygen in helium at 653 K. Surface coverage, $\Phi_{Fe\ atoms}$, is additionally shown. Acro: acrolein, Acet: acetaldehyde.

	Iron loading / wt%	Selectivity / %				$\Phi_{Fe\ atoms}$ / Fe atoms nm⁻²
		Acro	Acet	CO	CO_2	
Citrate	2.5	24.7	8.2	20.7	44.4	0.37
	3.0	29.1	8.5	21.0	39.1	0.47
	5.5	23.0	5.5	23.5	46.7	0.88
	6.3	23.5	7.8	21.3	45.6	1.04
	10.7	12.9	3.4	27.0	55.6	1.90
Nitrate	2.0	11.4	3.8	27.1	56.4	0.30
	3.3	20.0	3.6	25.1	50.3	0.52
	5.8	15.1	2.7	25.8	55.6	0.95
	7.2	20.1	4.2	23.3	51.3	1.19
	9.3	20.4	3.0	22.5	53.2	1.58
Fe_2O_3/SBA-15	10.5	0.5	0.2	4.5	94.7	

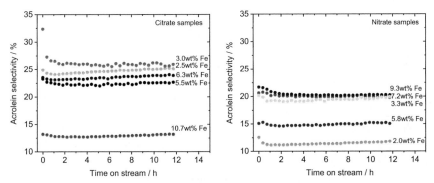

Figure 5.44: Acrolein selectivity as function of time on stream for citrate samples (left) and nitrate samples (right) during selective oxidation of propene (5% propene, 5% oxygen in helium, 653 K).

Average acrolein selectivity during 12 h time on stream in selective oxidation of propene as function of iron loading is depicted in Figure 5.45. A volcano-type behavior was observed. For iron loadings lower than 4.0wt%, acrolein selectivity sharply increased with increasing iron loading. At iron loadings higher than 4.0wt%, increasing iron loading was accompanied by decreasing acrolein selectivity. The decrease in acrolein selectivity with increasing iron loading was less pronounced than the increase for iron loadings lower than 4.0wt%. The maximum acrolein selectivity was observed for 3.0wt% Fe_Citrate.

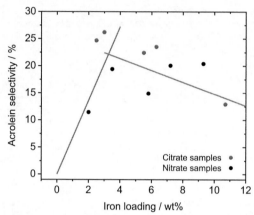

Figure 5.45: Acrolein selectivity as function of iron loading for Fe$_x$O$_y$/SBA-15 catalysts during selective oxidation of propene (5% propene, 5% oxygen in helium at 653 K).

To further investigate the influence of both precursor and iron loading on catalytic performance of Fe$_x$O$_y$/SBA-15 samples, acrolein selectivity and reaction rate of propene conversion as function of iron loading are compared in Figure 5.46. Citrate samples showed higher reaction rates of propene conversion compared to nitrate samples with similar iron loading. This was indicative of higher catalytic activity of the citrate samples. Apparently, smaller iron oxidic species and significantly higher amounts of β-FeOOH-like species of the citrate samples induced an increased catalytic activity. Selectivity towards desired reaction product acrolein followed the same trend for iron loadings up to 7.2wt%. Within the citrate samples, no significant trend in reaction rate of propene conversion as function of iron loading was observed. Conversely, within the nitrate samples, reaction rate of propene conversion decreased with increasing iron loading up to 7.2wt%. After that point, further increasing iron loading induced a slightly increased reaction rate. Accordingly, for nitrate samples with iron loadings up to 7.2wt%, larger iron oxidic species correlated with a decreasing catalytic activity in selective oxidation of propene. However, highest loaded nitrate sample, consisting of largest iron oxidic species showed an increased reaction rate compared to that of 7.2wt% Fe_Nitrate and highest acrolein selectivity. This might be ascribed to lowest degree of micropore filling and roughest SBA-15 surface for this sample. For Fe$_x$O$_y$/SBA-15 samples, iron loading-dependent particle size effect seemed not to be merely responsible for observed catalytic performance. Independent of the precursor, lowest iron loadings, i.e. 2.0wt% Fe_Nitrate and 2.5wt% Fe_Citrate, correlated with highest reaction rate of propene conversion, and hence, highest catalytic activity. This might be ascribed to highest amount of β-FeOOH-like species rather than to lowest Fe_xO_y sizes.

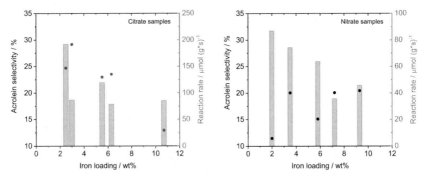

Figure 5.46: Acrolein selectivity (left axis, dots) and reaction rate of propene conversion (right axis, bars) for citrate samples (left) and nitrate samples (right) as function of iron loading. Measurements were performed in 5% propene, 5% oxygen in helium at 653 K.

All Fe_xO_y/SBA-15 samples, showed a selectivity towards CO_2 which was approximately twice as high as that towards CO (Figure 5.47). Once again, an opposite behavior of citrate and nitrate samples was revealed. For citrate samples, highest selectivities towards CO and CO_2 resulted for highest iron loading. Conversely, for nitrate samples, highest selectivities towards CO and CO_2 were observed for lowest iron loading. Interestingly, highest values of selectivities towards CO and CO_2 for sample 10.7wt% Fe_Citrate and 2.0wt% Fe_Nitrate were similar.

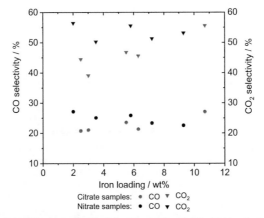

Figure 5.47: Selectivity towards CO (left axis, dots) and CO_2 (right axis, triangles) as function of iron loading for citrate samples (orange) and nitrate samples (blue) in selective oxidation of propene (5% propene, 5% oxygen in helium at 653 K).

Besides acrolein and carbon oxides, acetaldehyde was a main reaction product in selective oxidation of propene using Fe_xO_y/SBA-15 catalysts. Figure 5.48 depicts evolution of acetaldehyde selectivity as function of iron loading with respect to the precursor. Accordingly, the citrate precursor yielded significantly higher values of acetaldehyde selectivity compared to the nitrate precursor.

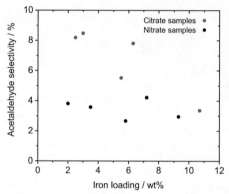

Figure 5.48: Selectivity towards acetaldehyde as function of iron loading in selective oxidation of propene (5% propene, 5% oxygen in helium at 653 K) for Fe_xO_y/SBA-15 catalysts.

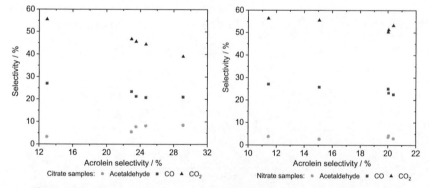

Figure 5.49: Selectivity towards acetaldehyde, CO, and CO_2 as function of acrolein selectivity for citrate samples (left) and nitrate samples (right). Selectivities were determined for selective oxidation of propene (5% propene, 5% oxygen in helium at 653 K).

Within the citrate samples, increasing acrolein selectivity correlated with an increasing selectivity towards acetaldehyde at the expense of CO and CO_2 (Figure 5.49, left and Table 5.13). Conversely, for the nitrate samples, no correlation between acrolein selectivity and selectivity towards acetaldehyde was revealed (Figure 5.49, right). Variations in selectivity towards acetaldehyde, CO, and CO_2 dependent on acrolein selectivity were less pronounced for nitrate samples, while citrate samples showed more pronounced variations. These precursor-dependent differences in product distributions might be indicative of different degrees of consecutive degradation and total oxidation reaction. In selective oxidation of propene, carbon oxides are mainly ascribed to total oxidation reaction. Conversely, acetaldehyde as by-product presumably results from consecutive oxidation of propionaldehyde being one of three possibly formed aldehydes during selective oxidation of propene. It is assumed that selective oxidation of propene starts with formation of three possible alcohol intermediates, allylalcohol, n-propanol, and isopropanol. These alcohols are afterwards oxidized towards corresponding aldehydes, acrolein, propionaldehyde, and acetone. Possible consecutive oxidation of the aldehydes might lead to formation of acrylic acid, acetaldehyde, and acetic acid. [10, 168] Based on this reaction mechanism, it seemed likely that the nitrate precursor induced a higher degree of total oxidation reaction compared to the citrate precursor. A favored selective oxidation of the citrate samples coincided with less total oxidation reaction. However, higher selectivity towards acetaldehyde for the citrate samples indicated a higher degree of consecutive oxidation during selective oxidation of propene. These differences in degrees of consecutive degradation and total oxidation dependent on precursor might be ascribed to precursor-dependent differences in iron oxidic species. Furthermore, precursor-dependent differences in reducibility and solid-state kinetic properties of Fe_xO_y/SBA-15 catalysts might also be responsible for these differences.

5.5.4   Structural evolution under catalytic conditions

In situ XAS during selective oxidation of propene

Fe_xO_y/SBA-15 catalysts were investigated by *in situ* X-ray absorption spectroscopy during selective oxidation of propene. Fe K edge X-ray absorption near edge structure (XANES) spectra of samples 6.3wt% Fe_Citrate and 7.2wt% Fe_Nitrate under catalytic conditions are depicted in Figure 5.50 and Figure 5.51. Fe K edge XANES spectra of the other Fe_xO_y/SBA-15 catalysts are given in the Appendix (Figure A.12.3 - Figure A.12.6). Evolution of Fe K edge XANES spectra showed slight variations in overall shape with increasing temperature which might be ascribed to variations in phase composition of Fe_xO_y/SBA-15 catalysts.

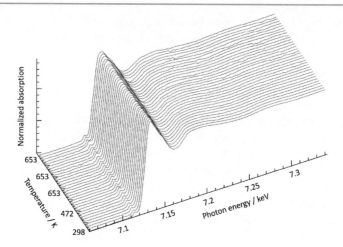

Figure 5.50: *In situ* XANES spectra of 6.3wt% Fe_Citrate during selective oxidation of propene. (5% propene, 5% oxygen in helium, temperature range 298–653 K, heating rate 5 K/min).

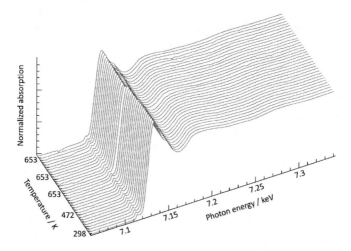

Figure 5.51: *In situ* XANES spectra of 7.2wt% Fe_Nitrate during selective oxidation of propene. (5% propene, 5% oxygen in helium, temperature range 298–653 K, heating rate 5 K/min).

LCXANES analysis during selective oxidation of propene

Linear combination of α-Fe_2O_3 and β-FeOOH XANES reference spectra was used to simulate experimental XANES spectra of Fe_xO_y/SBA-15 catalysts. Subsequently, evolution of phase composition of Fe_xO_y/SBA-15 during selective oxidation of propene was investigated. Quantification of α-Fe_2O_3-like and β-FeOOH-like species during selective oxidation of propene for lowest, middle, and highest loaded citrate sample are depicted in Figure 5.52. For all citrate samples, an increasing amount of β-FeOOH-like species at the expense of α-Fe$_2$O$_3$-like species with increasing temperature was revealed. However, evolution of phase composition during selective oxidation of propene significantly differed dependent on iron loading. Lowest loaded citrate sample, 2.5wt% Fe_Citrate, showed the fastest rise in amount of β-FeOOH-like species with increasing temperature. The amount of β-FeOOH-like species exceeded 50% at 429 K and further increased up to 90% at 600 K. Increasing the temperature up to 653 K induced no further increase in amount of β-FeOOH-like species. After 10.5 min holding time at 653 K amount of β-FeOOH-like species started to decrease. Higher loaded citrate samples differed not only in amount of β-FeOOH-like species but also in temperature of maximum amount of β-FeOOH-like species. 6.3wt% Fe_Citrate showed a slower increase in amount of β-FeOOH-like species with increasing temperature during selective oxidation of propene. Amount of 50% of β-FeOOH-like species was exceeded at 472 K. A constant amount of β-FeOOH-like species was reached at 582 K. Holding time at 653 K did not significantly influence the amount of β-FeOOH-like species for 6.3wt% Fe_Citrate. Highest loaded citrate sample, 10.7wt% Fe_Citrate, showed a similar evolution of phase composition during selective oxidation of propene as 6.3wt% Fe_Citrate. The increase in amount of β-FeOOH-like species with increasing temperature proceeded more slowly so that 50% β-FeOOH-like species was exceeded at 523 K. Reaching the catalysis temperature of 653 K induced a slight jump in amount of β-FeOOH-like species from 60.5% to 67.4%. Afterwards, amount of β-FeOOH-like species was constant and started to decrease after 7 min holding time.

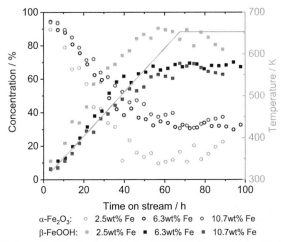

α-Fe$_2$O$_3$:   ○ 2.5wt% Fe ○ 6.3wt% Fe ○ 10.7wt% Fe
β-FeOOH: ■ 2.5wt% Fe ■ 6.3wt% Fe ■ 10.7wt% Fe

Figure 5.52: Quantification of α-Fe$_2$O$_3$-like and β-FeOOH-like species for citrate samples during selective oxidation of propene (5% propene, 5% oxygen in helium, temperature range 298-653 K, heating rate 5 K/min).

Evolution of phase composition of nitrate samples (Figure 5.53) during selective oxidation of propene also showed significant differences dependent on iron loading. As already observed for citrate samples, lowest loaded nitrate sample exhibited the fastest increase in amount of β-FeOOH-like species with increasing temperature. Reaching a temperature of 506 K correlated with an amount of β-FeOOH-like species higher than 50% for 2.0wt% Fe_Nitrate. At 523 K the maximum amount of β-FeOOH-like species of 63% was observed. During holding time at 653 K, values fluctuated which was ascribed to the data quality of this lowest loaded sample. Therefore, amount of β-FeOOH-like species was considered as invariant at approximately 63%. For higher loaded nitrate samples, the increase in amount of β-FeOOH-like species with increasing temperature proceeded slower. An amount of β-FeOOH-like species of more than 50% was observed for 7.2wt% Fe_Nitrate when reaching the catalysis temperature of 653 K. During holding time at 653 K, amount of β-FeOOH-like species slightly fluctuated around 50-52%. Highest loaded nitrate sample showed a slightly slower increase in amount of β-FeOOH-like species with increasing temperature. Reaching the holding time at 653 K correlated with the maximum amount of β-FeOOH-like species of 32% for sample 9.3wt% Fe_Nitrate. Afterwards, values slightly fluctuated around 32%, and decreased after 20 min holding time to 28%.

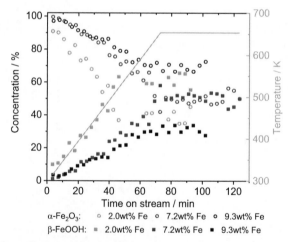

Figure 5.53: Quantification of α-Fe$_2$O$_3$-like and β-FeOOH-like species for nitrate samples during selective oxidation of propene (5% propene, 5% oxygen in helium, temperature range 298-653 K, heating rate 5 K/min).

LCXANES analysis during selective oxidation of propene revealed a correlation between increasing temperature and increasing amount of β-FeOOH-like species for Fe$_x$O$_y$/SBA-15 catalysts independent of both precursor and iron loading. However, magnitude of the increase in amount of β-FeOOH-like species under catalytic conditions was strongly affected by iron loading and precursor. Independent of the precursor, lowest loaded samples showed both fastest increase and highest amount of β-FeOOH-like species with increasing temperature. Increasing the iron loading induced a slower increase in amount of β-FeOOH-like species with increasing temperature and lower amounts of β-FeOOH-like species. For Fe_xO_y/SBA-15 catalysts obtained from citrate precursor, the amount of β-FeOOH-like species exceeded 50% at lower temperatures compared to corresponding nitrate samples. Furthermore, citrate samples possessed significantly higher amounts of β-FeOOH-like species during selective oxidation of propene compared to the nitrate samples. Accordingly, for citrate samples with already higher amounts of β-FeOOH-like species before catalytic reaction, formation of additional β-FeOOH-like species was favored.

Local structure of Fe_xO_y/SBA-15 after catalytic reaction
Fe K edge X-ray absorption near edge structure spectra of Fe_xO_y/SBA-15 samples before and after selective oxidation of propene are depicted in Figure 5.54. XANES spectra of nitrate samples (Figure 5.54, right) before and after catalytic reaction showed a similar overall shape. For highest loaded nitrate sample, recording XANES spectra after selective oxidation of propene was not possible. Nitrate samples after catalytic reaction showed XANES spectra resembling β-FeOOH reference spectrum (Figure 5.18, left). Furthermore, small pre-edge peak was retained. Citrate samples showed similar variations in XANES spectra before and after catalytic reaction (Figure 5.54, left). Additionally, absorption beyond the absorption edge became broader. XANES spectra of citrate samples appeared to be more affected by treatment under catalytic reaction conditions compared to those of nitrate samples. These slight differences in XANES spectra before and after selective oxidation of propene might be indicative of differences in local structure of Fe_xO_y/SBA-15 samples resulted from treatment under catalytic reaction conditions.

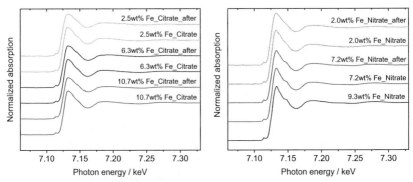

Figure 5.54: Fe K edge XANES spectra before and after selective oxidation of propene (5% propene, 5% oxygen in helium at 653 K) for citrate (left) and nitrate (right) samples. Spectra were recorded at ambient temperature in helium.

For determining phase composition of Fe_xO_y/SBA-15 samples after catalytic reaction, experimental XANES spectra were simulated by fitting various linear combinations of iron oxide reference spectra. For all Fe_xO_y/SBA-15 samples, only combination of two different iron oxide reference spectra yielded satisfying fit results. Required iron oxide references were α-Fe_2O_3 and β-FeOOH. Hence, phase composition of Fe_xO_y/SBA-15 samples remained unchanged. Fe_xO_y/SBA-15 samples still consisted of different iron oxidic species with a local geometry similar to that of α-Fe_2O_3 or β-FeOOH. However, amount of α-Fe_2O_3-like and β-FeOOH-like species was significantly affected by treatment under catalytic reaction conditions (Table 5.14). For all Fe_xO_y/SBA-15 samples, an increased amount of β-FeOOH-like species after catalytic reaction was revealed. The citrate precursor induced an amount of β-FeOOH-like species of at least 46.7% after catalytic reaction. Conversely, for nitrate samples, β-FeOOH-like species remained distinctly in the minority. LCXANES analysis clarified precursor-dependent and iron loading-dependent differences in phase composition of Fe_xO_y/SBA-15 samples after selective oxidation of propene. Accordingly, treatment under catalytic reaction conditions did not influence the trend in phase composition of Fe_xO_y/SBA-15 samples dependent on iron loading and precursor (see chapter 5.3.4). However, amount of β-FeOOH-like species in Fe_xO_y/SBA-15 samples was distinctly affected by treatment under catalytic reaction conditions. This complied with the variations in Fe K edge XANES spectra of Fe_xO_y/SBA-15 samples before and after catalytic reaction (Figure 5.54).

Table 5.14: Fraction of α-Fe_2O_3- and β-FeOOH-like species in Fe_xO_y/SBA-15 before and after selective oxidation of propene (5% propene, 5% oxygen in helium, 653 K) resulted from phase composition analysis of Fe K edge XANES spectra. XANES spectra were recorded at ambient temperature in helium.

	After catalytic reaction		Before catalytic reaction	
	α-Fe_2O_3 / %	β-FeOOH / %	α-Fe_2O_3 / %	β-FeOOH / %
2.5wt% Fe_Citrate	32.3	67.7	74.8	25.2
6.3wt% Fe_Citrate	52.3	47.7	87.0	13.0
10.7wt% Fe_Citrate	53.3	46.7	91.8	8.2
2.0wt% Fe_Nitrate	60.9	39.1	84.7	15.3
7.2wt% Fe_Nitrate	76.5	23.5	97.6	2.4
9.3wt% Fe_Nitrate	-	-	99.9	0.1

In addition to XANES analysis, extended X-ray absorption fine structure of the Fe_xO_y/SBA-15 samples after selective oxidation of propene was analyzed. For elucidating the influence of treatment under catalytic reaction conditions on local structure of iron oxidic species on SBA-15, results from EXAFS analysis before and after catalytic reaction were compared. The pseudo radial distribution function (not phase shift corrected), $FT(\chi(k)*k^3)$, of Fe_xO_y/SBA-15 samples before and after catalytic reaction is depicted in Figure 5.55. Only minor differences in amplitude and overall shape of the $FT(\chi(k)*k^3)$ were observed. Independent of the precursor, after catalytic reaction, the $FT(\chi(k)*k^3)$ showed a minor decrease in amplitude of both peaks corresponding to Fe-O and Fe-Fe shells. Furthermore, the second peak in the pseudo radial distribution function of all citrate samples after catalytic reaction was shifted to lower distances. A similar shift was not observed for the nitrate samples. As already described for Fe_xO_y/SBA-15 samples before catalytic reaction (see chapter 5.3.4), the amplitude of both first and second peak in the $FT(\chi(k)*k^3)$ of the citrate samples was lower compared to that of nitrate samples after catalytic reaction.

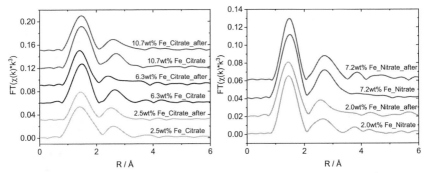

Figure 5.55: Pseudo radial distribution function (not phase shift corrected), $FT(\chi(k)*k^3)$, of Fe_xO_y/SBA-15 samples before and after selective oxidation of propene (5% propene, 5% oxygen in helium at 653 K) with an offset in y-axis. EXAFS spectra were recorded at ambient temperature in helium.

EXAFS refinements for the Fe_xO_y/SBA-15 samples after catalytic reaction were performed for a more detailed structural analysis. Therefore, theoretical X-ray absorption fine structure calculated for a α-Fe_2O_3 model structure was refined to the experimental $FT(\chi(k)*k^3)$ of Fe_xO_y/SBA-15 samples. Fitting procedure was conducted analogously as described in chapter 5.3.4. A good agreement between theoretical and experimental $FT(\chi(k)*k^3)$ of the Fe_xO_y/SBA-15 samples was achieved by using α-Fe_2O_3 as structural model. This indicated that edge-, corner-, and face-shared $[FeO_6]$ structural motifs of the model structure were still sufficient for describing the experimental $FT(\chi(k)*k^3)$. Theoretical and experimental $FT(\chi(k)*k^3)$ are shown in Figure 5.56.

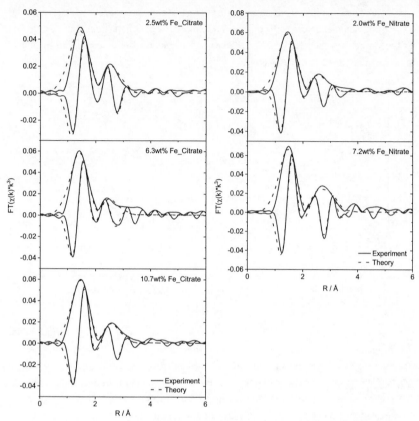

Figure 5.56: Experimental (straight line) and theoretical (dotted line) Fe K edge FT(χ(k)*k^3) of citrate (left) and nitrate (right) samples after catalytic reaction.

Table 5.15: EXAFS fit results for Fe_xO_y/SBA-15 samples after selective oxidation of propene. Type of neighbor, coordination number, N, and XAFS disorder parameter, σ^2, of atoms at the distance R from the Fe atoms in Fe_xO_y/SBA-15 samples at various iron loadings. Experimental parameters were obtained from refinement of α-Fe_2O_3 model structure (AMCSD 0017806 [138]) to the experimental Fe K edge $FT(\chi(k)*k^3)$ of Fe_xO_y/SBA-15 (Figure 5.56). Fit range in R space 1.1-3.0 Å (or 1.1-3.4 Å); k space 2.2-10.3 Å^{-1} (or 2.2-9.2 Å^{-1}); N_{ind} = 11.4 (or 10.1); N_{free} = 5; Number of SS paths = 3 (or 2); S_0^2 = 0.9. Alternative values of fitting parameters given in parentheses refer to slightly modified refinement for lowest loaded citrate and nitrate sample due to $\chi(k)*k^3$ data quality.

	Type	N_{fixed}	R / Å	σ^2 / Å^2	E_0 / eV	Fit residual R
2.5wt% Fe_Citrate	Fe-O	6	1.92	0.0204	+0.18	6.0
	Fe-Fe	3	2.99	0.0167		
	Fe-Fe	3	-	-		
6.3wt% Fe_Citrate	Fe-O	6	1.91	0.0165	-0.52	5.9
	Fe-Fe	3	2.96	0.0219		
	Fe-Fe	3	-	-		
10.7wt% Fe_Citrate	Fe-O	6	1.94	0.0166	+1.81	4.8
	Fe-Fe	3	3.01	0.0190		
	Fe-Fe	3	-	-		
2.0wt% Fe_Nitrate	Fe-O	6	1.94	0.0165	+1.24	5.5
	Fe-Fe	3	2.99	0.0194		
	Fe-Fe	3	-	-		
7.2wt% Fe_Nitrate	Fe-O	6	1.94	0.0143	+0.51	10.4
	Fe-Fe	3	3.00	0.0183		
	Fe-Fe	3	3.42	0.0172		

For all Fe_xO_y/SBA-15 samples, the coordination number for the Fe-O backscattering path was kept invariant at six. As explained in chapter 5.3.4, variations in the amplitude in the $FT(\chi(k)*k^3)$ of Fe_xO_y/SBA-15 samples might depend on coordination number, static disorder, or variations in phase composition. Hence, analogously to variations in the amplitude in the $FT(\chi(k)*k^3)$ of Fe_xO_y/SBA-15 samples before catalytic reaction, those after catalytic reaction were ascribed to variations in the phase composition. Consequently, these variations in the $FT(\chi(k)*k^3)$ induced the variations in the Debye-Waller factors, $\sigma^2(Fe-O)$ (Figure 5.57, left). All Fe_xO_y/SBA-15 samples showed an increased $\sigma^2(Fe-O)$ after catalytic reaction. $\sigma^2(Fe-O)$ of Fe_xO_y/SBA-15 samples after catalytic reaction as function of iron loading showed the same trend as that observed before catalytic reaction. Independent of the precursor, increasing iron loading correlated with decreasing $\sigma^2(Fe-O)$. Additionally, $\sigma^2(Fe-O)$ of the citrate samples were significantly higher than those of the nitrate samples before and after catalytic reaction. Hence, treatment under catalytic reaction conditions did merely influence values of $\sigma^2(Fe-O)$ of Fe_xO_y/SBA-15 samples, while precursor-dependent and iron loading-dependent trends of $\sigma^2(Fe-O)$ remained unaffected. Accordingly, $\sigma^2(Fe-O)$ was again assigned to the phase fraction of β-FeOOH-like species, and hence, to the presence of both Fe-O and Fe-OH bonds in the

Fe$_x$O$_y$/SBA-15 samples. Treatment under catalytic reaction conditions induced an increased amount of β-FeOOH-like species in Fe$_x$O$_y$/SBA-15 samples. These differences in amount of β-FeOOH-like species were correlated with higher values of σ^2(Fe-O) after catalytic reaction. *R(Fe-O)* before and after catalytic reaction of the lowest loaded nitrate sample remained unaffected, while that of 7.2wt% Fe_Nitrate and those of the citrate samples slightly decreased after catalytic reaction (Figure 5.57, right). This slight decrease in *R(Fe-O)* might be indicative of an increased degree of aggregation of iron oxidic species after catalytic reaction.

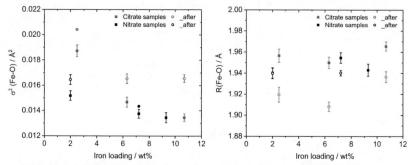

Figure 5.57: Fe-O Debye-Waller factor (XAFS disorder parameter), σ^2 (left), and Fe-O distance, R (right), of Fe$_x$O$_y$/SBA-15 samples before (filled squares) and after (empty dots) catalytic reaction as function of iron loading. Parameters resulted from refinements of Fe K edge FT(χ(k)*k^3) (Table 5.15).

Significant differences induced from treatment under catalytic reaction conditions were also revealed in Debye-Waller factor, σ^2*(Fe-Fe_edge)*, of edge-shared [FeO$_6$] octahedra of Fe$_x$O$_y$/SBA-15 samples (Figure 5.58, left). Independent of the precursor, σ^2*(Fe-Fe_edge)* were significantly increased after catalytic reaction, except for lowest loaded citrate and nitrate sample. As observed for σ^2*(Fe-O)*, the behavior of σ^2*(Fe-Fe_edge)* as function of iron loading remained unaffected by treatment under catalytic reaction conditions. Additionally, precursor-dependent differences in σ^2*(Fe-Fe_edge)* were retained after catalytic reaction. However, treatment under catalytic reaction conditions induced an increase in σ^2*(Fe-Fe_edge)* for iron loadings higher than 2.5wt% independent of precursor. For Fe$_x$O$_y$/SBA-15 samples after catalytic reaction, the same assumptions for the Fe-Fe shell were made as before catalytic reaction (see chapter 5.3.4). Hence, assuming the amount of β-FeOOH-like species slightly affected σ^2*(Fe-Fe_edge)* and *R(Fe-Fe_edge)*, two superimposed effects had to be considered for the Fe-Fe shell. For Fe$_x$O$_y$/SBA-15 samples before catalytic reaction, the influence of average coordination number was considered highly relevant for the Fe-Fe shell, while that of phase composition was considered minor. Based on significant variations in phase composition of Fe$_x$O$_y$/SBA-15 samples after catalytic reaction, both effects, average coordination number and phase composition, seemed now equally relevant for the Fe-Fe shell. The average coordination number for the Fe-Fe backscattering path was set to three and kept invariant within the EXAFS refinements. For Fe$_x$O$_y$/SBA-15 samples, average coordination number after catalytic reaction was presumably lower than three. Therefore, the decrease of the amplitude in the FT(χ(k)*k^3) induced an increased σ^2(Fe-Fe_edge) after catalytic reaction. Higher values of σ^2(Fe-Fe_edge) for citrate samples after catalytic reaction compared to those of the nitrate samples still agreed with smaller, less aggregated iron oxidic species and higher amounts of β-FeOOH-like species obtained from citrate precursor. Differences in R(Fe-Fe_edge) for Fe$_x$O$_y$/SBA-15 samples after catalytic reaction might also be ascribed to variations in average coordination number and phase composition. For nitrate samples, R(Fe-Fe_edge) were largely independent of iron loading before and after catalytic reaction. Conversely, R(Fe-Fe_edge) of citrate samples were affected by treatment under catalytic reaction conditions (Figure 5.58, right). The decreased R(Fe-Fe_edge) may indicate an increased degree of aggregation of iron oxidic species after catalytic reaction.

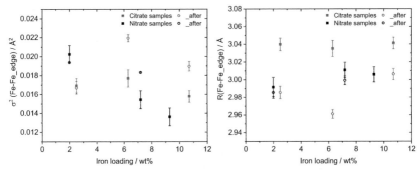

Figure 5.58: Fe-Fe Debye-Waller factor (XAFS disorder parameter), σ^2 (left), and Fe-Fe distance, R (right), of Fe$_x$O$_y$/SBA-15 samples before (filled squares) and after (empty dots) catalytic reaction as function of iron loading. Parameters resulted from refinements of Fe K edge FT(χ(k)*k^3) (Table 5.15).

For 7.2wt% Fe_Nitrate after catalytic reaction, an additional Fe-Fe backscattering path ascribed to Fe-Fe in corner-shared [FeO$_6$] octahedra was still required to fully account for the amplitude in the FT(χ(k)*k^3). Average coordination number of this backscattering path was set to three and kept invariant in EXAFS refinement. However, a deviation of an average coordination number of three seemed also reasonable for 7.2wt% Fe_Nitrate after catalytic reaction, as assumed before catalytic reaction (see chapter 5.3.4). For this sample, deviation of an average coordination number of three was decreased by treatment under catalytic reaction conditions. This presumably induced the decreased σ^2(Fe-Fe_corner) of 7.2wt% Fe_Nitrate after catalytic reaction (Table 5.15). Fe-Fe distance, R(Fe-Fe_corner), remained unaffected by treatment under catalytic reaction conditions.

Consequently, treatment under catalytic reaction conditions particularly affected the phase composition of Fe$_x$O$_y$/SBA-15 samples, while Fe-O and Fe-Fe distances remained less affected. The citrate samples favored formation of β-FeOOH-like species. Independent of the precursor, treatment under catalytic reaction conditions induced an increased amount of β-FeOOH-like species. Furthermore, a slightly increased degree of aggregation after catalytic reaction, especially for citrate samples, seemed likely.

5.5.5 Influence of β-FeOOH-like species on catalytic performance

Differences in phase composition of Fe$_x$O$_y$/SBA-15 catalysts dependent on precursor and iron loading might have a crucial impact on catalytic performance. To further elucidate a possible influence of β-FeOOH-like species on catalytic performance of Fe$_x$O$_y$/SBA-15, correlation between amount of β-FeOOH-like species and acrolein selectivity during selective oxidation of propene was investigated. Therefore, average acrolein selectivity and average concentration of β-FeOOH-like species during holding time at 653 K as function of iron loading are depicted in Figure 5.59. For citrate samples (Figure 5.59, left), increasing average concentration of β-FeOOH-like species with decreasing iron loading correlated with an increasing acrolein selectivity. In contrast to citrate samples, nitrate samples showed a correlation between increasing average concentration of β-FeOOH-like species with decreasing iron loading and decreasing acrolein selectivity (Figure 5.59, right). Due to this opposite trend, a further effect was presumably accounting for selectivity towards acrolein.

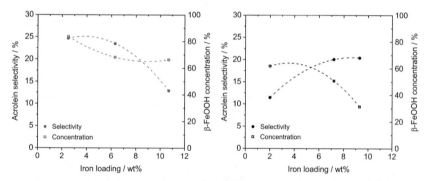

Figure 5.59: Average acrolein selectivity (left axis) and average concentration of β-FeOOH-like species (right axis) during holding time at 653 K in 5% propene, 5% oxygen in helium. Left: Citrate samples. Right: Nitrate samples. Dotted lines only for visualization.

This different influence of β-FeOOH-like species on catalytic performance dependent on precursor was furthermore reflected by onset of catalytic activity expressed by acrolein ion current. For 10.7wt% Fe_Citrate, exceeding the amount of β-FeOOH-like species of 50% was accompanied by an increase in acrolein ion current (Figure 5.60). The slight decrease in amount of β-FeOOH-like species during holding time at 653 K was accompanied by a slight decrease in acrolein ion current. This was indicative of β-FeOOH-like species accounting for the selective oxidation of propene to acrolein. Conversely, for nitrate samples, a direct correlation between amount of β-FeOOH-like species and onset of acrolein ion current was not observed (Figure 5.61). Consequently, besides amount of β-FeOOH-like species in

Fe$_x$O$_y$/SBA-15 catalysts, differences in reducibility and surface and porosity characteristics presumably determined catalytic performance, particularly for the nitrate samples.

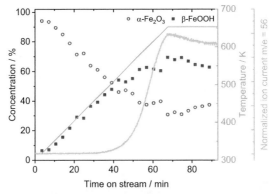

Figure 5.60: Evolution of amount of α-Fe$_2$O$_3$-like and β-FeOOH-like species for 10.7wt% Fe_Citrate during selective oxidation of propene (5% propene, 5% oxygen in helium, temperature range 298-653 K, heating rate 5 K/min) together with normalized ion current of acrolein (m/e = 56). Ion current was normalized to that of helium.

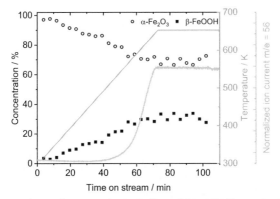

Figure 5.61: Evolution of amount of α-Fe$_2$O$_3$-like and β-FeOOH-like species for 9.3wt% Fe_Citrate during selective oxidation of propene (5% propene, 5% oxygen in helium, temperature range 298-653 K, heating rate 5 K/min) together with normalized ion current of acrolein (m/e = 56). Ion current was normalized to that of helium.

5.5.6 Influence of Fe$_x$O$_y$ reducibility on catalytic performance

For selective oxidation of propene, an intermediate Fe-O bond strength is required to obtain selective oxidation products. Too weak Fe-O bonds will lead to predominantly total oxidation products, whereas too strong Fe-O bonds will inhibit the reaction. [7] A significant influence of reducibility of supported metal oxide catalysts on their catalytic performance in selective oxidation reactions is reported in literature concerning the so-called Mars-van-Krevelen mechanism. [17, 48, 50] Therefore, influence of reducibility of Fe_xO_y species on their catalytic performance was investigated. Figure 5.62 depicts correlation between acrolein selectivity in selective oxidation of propene and apparent activation energy of rate-determining step in reduction determined by Kissinger method for Fe_xO_y/SBA-15 catalysts. For nitrate samples (Figure 5.62, right), acrolein selectivity increased with decreasing apparent activation energy in reduction. Accordingly, better reducibility of larger iron oxidic species with increasing iron loading induced a higher selectivity towards desired reaction product acrolein. For citrate samples, correlation between acrolein selectivity and reducibility appeared more complex (Figure 5.62, left). Apparent activation energy in reduction for highest loaded citrate samples was constant and only decreased with decreasing iron loading from 6.3wt%. Conversely, acrolein selectivity increased with decreasing iron loading from 10.7 through 3.0wt%. Lowest loaded citrate sample, 2.5wt% Fe_Citrate, possessed lowest apparent activation energy and a slightly decreased acrolein selectivity compared to 3.0wt% Fe_Citrate. These observations suggested reducibility being not the only essential factor for selectivity in selective oxidation of propene. For Fe_xO_y/SBA-15 catalysts obtained from citrate precursor, amount of β-FeOOH-like species (see chapter 5.5.5) was considered important besides reducibility.

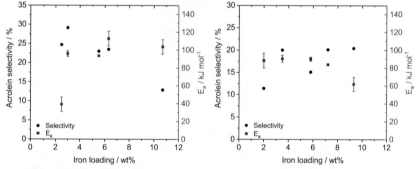

Figure 5.62: Correlation between acrolein selectivity (left axis) and apparent activation energy of rate-determining step in reduction determined by Kissinger method (right axis) as function of iron loading. Acrolein selectivity was measured in 5% propene, 5% oxygen in helium at 653 K and reduction was conducted in 5% H_2 in argon up to 1223 K. Left: Citrate samples. Right: Nitrate samples.

Reducibility, i.e. oxygen availability, of Fe_xO_y/SBA-15 catalysts considerably influenced catalytic performance. The nitrate precursor induced larger and higher aggregated Fe_xO_y species on SBA-15 compared to the citrate precursor. Increasing the iron loading within the nitrate samples resulted in a rougher SBA-15 surface and a higher contribution of micropores to entire surface area. This indicated less pronounced interactions between Fe_xO_y species and support material for higher iron loadings resulting in a better reducibility. The increased oxygen availability of higher loaded nitrate samples consequently induced the higher acrolein selectivity. Conversely, for citrate samples, highest contribution of micropores to entire surface area together with roughest SBA-15 surface was observed for lowest iron loading. Accordingly, decreasing iron loading within the citrate samples yielded not only higher amounts of β-FeOOH-like species but also less pronounced interactions between Fe_xO_y species and support material. These species possessing a better reducibility and increased oxygen availability were associated with a higher selectivity towards acrolein. The observed influence of reducibility of Fe_xO_y species on catalytic performance consequently complied with the results from structural characterization (see chapter 5.3.1).

5.5.7 A comparison of catalytic performance dependent on precursor

Previous chapters revealed distinct differences in catalytic performance of Fe_xO_y/SBA-15 catalysts dependent on both iron loading and precursor. To gain further insight into precursor-dependent effects on catalytic performance, samples 6.3wt% Fe_Citrate and 7.2wt% Fe_Nitrate with similar surface coverages (1.04 Fe atoms nm^{-2} and 1.19 Fe atoms nm^{-2}) were directly compared during selective oxidation of propene at 653 K. Evolution of reaction rate of propene conversion and acrolein formation rate during 12 h under propene oxidation conditions are depicted in Figure 5.63. Both reaction rate of propene conversion and acrolein formation rate of 6.3wt% Fe_Citrate was approximately three times higher than that of 7.2wt% Fe_Nitrate. This indicated a higher catalytic activity of 6.3wt% Fe_Citrate.

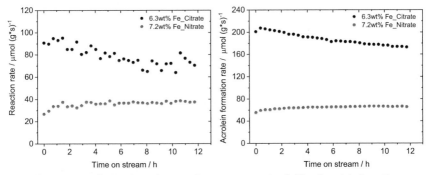

Figure 5.63: Evolution of reaction rate of propene conversion (left) and acrolein formation rate (right) as function of time on stream during selective oxidation of propene at 653 K (5% propene, 5% oxygen in helium) for 6.3wt% Fe_Citrate and 7.2wt% Fe_Nitrate.

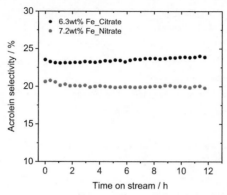

Figure 5.64: Acrolein selectivity as function of time on stream during selective oxidation of propene at 653 K (5% propene, 5% oxygen in helium) for 6.3wt% Fe_Citrate and 7.2wt% Fe_Nitrate.

Besides higher reaction rate of propene conversion and acrolein formation rate, 6.3wt% Fe_Citrate possessed a higher acrolein selectivity compared to 7.2wt% Fe_Nitrate (Figure 5.64). A similar behavior was observed by Torres Galvis et al. [42] investigating the effect of the precursor on catalytic performance of iron oxides supported on α-Al$_2$O$_3$ in Fischer-Tropsch synthesis. They observed a higher catalytic activity and a higher selectivity towards C$_2$-C$_4$ olefins by using a citrate precursor for iron loadings of 5 and 10wt% compared to a nitrate precursor. The reported precursor-dependent differences in catalytic performance were however merely attributed to differences in iron oxide distribution and aggregation. According to Tsoncheva et al. [56], variations in catalytic performance of iron oxide-based SBA-15 catalysts for methanol decomposition reaction resulted from a complex interplay of simultaneous effects. Particle aggregation, reductive phase transformation, and amount of micropores in SBA-15 were reported to be crucial for catalytic performance. Such a combination of simultaneous effects seemed also likely for Fe$_x$O$_y$/SBA-15 catalysts for selective oxidation of propene.

In situ LCXANES analysis during selective oxidation of propene revealed precursor-dependent differences in phase composition of Fe$_x$O$_y$/SBA-15 catalysts. Figure 5.65 compares evolution of amount of α-Fe$_2$O$_3$-like and β-FeOOH-like species as function of time on stream under propene oxidation conditions for samples 6.3wt% Fe_Citrate and 7.2wt% Fe_Nitrate. Sample 6.3wt% Fe_Citrate showed a significantly faster increase in amount of β-FeOOH-like species with increasing temperature yielding an invariant amount of β-FeOOH-like species of approximately 68% at 582 K. Conversely, 7.2wt% Fe_Nitrate showed a slower increase in amount of β-FeOOH-like species with increasing temperature. An invariant amount of approximately 50% β-FeOOH-like species was reached at 653 K. Differences in phase composition of Fe$_x$O$_y$/SBA-15 catalysts were ascribed to the chemical memory effect of Fe$_x$O$_y$/SBA-15 (see chapter 5.3.5). Consequently, chemical memory effect was not only observed after calcination but also during selective oxidation of propene.

β-FeOOH-like species being the majority compared to α-Fe$_2$O$_3$-like species seemed to be crucial for catalytic performance of Fe$_x$O$_y$/SBA-15 catalysts. 6.3wt% Fe_Citrate possessed not only smaller, higher dispersed iron oxidic species on SBA-15, but also significantly higher amounts of β-FeOOH-like species compared to 7.2wt% Fe_Nitrate. These precursor-dependent differences in phase composition together with the differences in iron oxidic species size and dispersion (see chapter 5.3.3) were considered responsible for the higher catalytic activity and acrolein selectivity of 6.3wt% Fe_Citrate compared to 7.2wt% Fe_Nitrate.

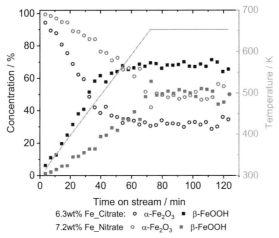

Figure 5.65: Quantification of α-Fe$_2$O$_3$-like and β-FeOOH-like species for 6.3wt% Fe_Citrate and 7.2wt% Fe_Nitrate during selective oxidation of propene (5% propene, 5% oxygen in helium at 653 K, temperature range 298-653 K, heating rate 5 K/min).

Besides amount of β-FeOOH-like species and surface and porosity characteristics, reducibility of Fe$_x$O$_y$/SBA-15 catalysts was revealed to have a severe impact on their catalytic performance. Figure 5.66 depicts evolution of normalized iron oxidation state of 6.3wt% Fe_Citrate and 7.2wt% Fe_Nitrate during selective oxidation of propene. Increasing temperature correlated with a slight reduction of Fe$_x$O$_y$/SBA-15 samples, independent of the precursor. However, 7.2wt% Fe_Nitrate showed a better reducibility (see chapter 5.4), and hence, a higher decrease in iron oxidation state. Increasing the temperature up to 653 K was accompanied by a decrease in iron oxidation state of 0.14 for 6.3wt% Fe_Citrate and 0.39 for 7.2wt% Fe_Nitrate. The higher degree of reduction for 7.2wt% Fe_Nitrate agreed with the lower apparent activation energy in reduction determined by Kissinger method. 6.3wt% Fe_Citrate showed no further reduction during time on stream at 653 K. Determined average iron oxidation state amounted to +2.85. Conversely, 7.2wt% Fe_Nitrate showed an increasing average iron oxidation state from +2.61 up to +2.71 during time on stream at 653 K. This increase in average iron oxidation state was indicative of a re-oxidation. However, despite proceeding re-oxidation, 7.2wt% Fe_Nitrate remained stronger reduced during time on stream at 653 K compared to 6.3wt% Fe_Citrate. Both catalysts remained in a partially reduced oxidation state under selective oxidation conditions.

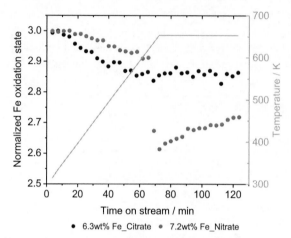

Figure 5.66: Normalized iron oxidation state of 6.3wt% Fe_Citrate and 7.2wt% Fe_Nitrate as function of time on stream during selective oxidation of propene (5% propene, 5% oxygen in helium at 653 K, temperature range 298-653 K, heating rate 5 K/min).

Table 5.16 summarizes results from *in situ* characterization of 6.3wt% Fe_Citrate and 7.2wt% Fe_Nitrate. Comparison of 6.3wt% Fe_Citrate and 7.2wt% Fe_Nitrate in selective oxidation of propene revealed two major effects determining the catalytic performance of Fe_xO_y/SBA-15 catalysts dependent on precursor. The first precursor-dependent effect was the amount of β-FeOOH-like species. 6.3wt% Fe_Citrate possessed higher amounts of β-FeOOH-like species, favoring formation of additional β-FeOOH-like species under propene oxidation conditions at 653 K. β-FeOOH-like species being the majority compared to α-Fe_2O_3-like species was directly correlated with the higher acrolein selectivity for 6.3wt% Fe_Citrate. The second precursor-dependent effect on catalytic performance of Fe_xO_y/SBA-15 catalysts was the reducibility of iron oxidic species. Lower apparent activation energy in reduction agreed with a higher difference in iron oxidation state for 7.2wt% Fe_Nitrate compared to 6.3wt% Fe_Citrate. Better reducibility of 7.2wt% Fe_Nitrate yielded a higher selectivity towards total oxidation products, CO and CO_2, at the expense of acrolein. Accordingly, larger iron oxidic species together with lower amounts of β-FeOOH-like species resulting for 7.2wt% Fe_Nitrate induced a better reducibility. Fe-O bond strength of supported iron oxidic species of 7.2wt% Fe_Nitrate seemed to be weaker compared to those of 6.3wt% Fe_Citrate, and hence, acrolein selectivity was lower for 7.2wt% Fe_Nitrate.

Apparently, at lower amounts of β-FeOOH-like species, reducibility was the dominating factor for catalytic performance. Comparison of 6.3wt% Fe_Citrate and 7.2wt% Fe_Nitrate suggested a combination of particle size effect, surface and porosity characteristics, phase composition, and reducibility determining the catalytic performance of Fe_xO_y/SBA-15 catalysts.

Table 5.16: Comparison of 6.3wt% Fe_Citrate and 7.2wt% Fe_Nitrate in selective oxidation of propene. Average selectivities determined during 12 h at 653 K in 5% propene, 5% oxygen in helium. Differences in average Fe oxidation state were determined under catalytic conditions while temperature was increased from 298 K to 653 K with a heating rate of 5 K/min. Apparent activation energy, E_a, during reduction (5% H_2 in argon at 1223 K) was determined by Kissinger method.

	6.3wt% Fe_Citrate	7.2wt% Fe_Nitrate
Average acrolein selectivity	23.5%	20.1%
Average amount of β-FeOOH-like species during time on stream at 653 K	68.1%	50.9%
E_a in reduction	113 ± 8 kJ/mol	84 ± 1 kJ/mol
Difference in average Fe oxidation state during ramp	0.14	0.39
Average selectivity towards carbon oxides	66.9%	74.6%

5.5.8 Correlation between catalytic performance, structural, and solid-state kinetic properties of Fe_xO_y/SBA-15 catalysts

Investigations of Fe_xO_y/SBA-15 catalysts under propene oxidation conditions revealed pronounced differences in catalytic performance dependent on precursor. Previous chapters suggested catalytic performance of supported iron oxidic species on SBA-15 being influenced by reducibility and solid-state kinetic properties, but also by structural properties. This chapter aims at elucidating a correlation between various effects and catalytic performance of Fe_xO_y/SBA-15 catalysts. *In situ* characterization of Fe_xO_y/SBA-15 catalysts indicated following effects determining catalytic performance: (I) particle size effect, (II) amount of β-FeOOH-like species, (III) degree of micropore filling and surface roughness as measure for interactions between iron oxidic species and support material, and (IV) reducibility.

Higher catalytic activity of citrate samples compared to nitrate samples with similar iron loading was ascribed to the first two effects. The citrate precursor induced significantly smaller and higher dispersed iron oxidic species accompanied by higher amounts of β-FeOOH-like species (see chapter 5.3.3 and 5.3.4). Furthermore, independent of the precursor, lowest loaded samples possessed highest reaction rate of propene conversion coinciding with smallest iron oxidic species and highest amount of β-FeOOH-like species.

For citrate samples, less formation of total oxidation products during selective oxidation of propene was observed compared to that of nitrate samples. Again, particle size effect seemed essential for this trend. Furthermore, reducibility, determined by degree of micropore filling, was crucial for the extend of total oxidation reaction. Citrate samples possessed a higher apparent activation energy of reduction compared to corresponding nitrate samples, except for 2.5wt% Fe_Citrate. This resulted in an intermediate oxygen availability during catalytic reaction, yielding less total oxidation products for the citrate samples. This trend was accompanied by a higher acrolein selectivity for citrate samples up to an iron loading of 7.2wt%. For Fe_xO_y/SBA-15 catalysts obtained from citrate precursor, amount of β-FeOOH-like species was considered essential for acrolein selectivity. Exceeding an amount of 50% of β-FeOOH-like species correlated with the onset of acrolein formation during selective oxidation of propene. Accordingly, the citrate precursor induced a higher amount of β-FeOOH-like species due to the chemical memory effect of Fe_xO_y/SBA-15 catalysts. This higher amount of β-FeOOH-like species favored formation of additional β-FeOOH-like species under catalytic conditions being directly correlated with catalytic performance. Increasing acrolein selectivity correlated with increasing amount of β-FeOOH-like species and decreasing iron loading for citrate samples. Hence, smaller iron oxidic species, and consequently, a lower degree of micropore filling for lower iron loadings together with the higher amount of β-FeOOH-like species within the citrate samples determined the catalytic performance in selective oxidation propene. Conversely, for nitrate samples, acrolein selectivity increased with increasing iron loading, and hence, with increasing iron oxidic species size and decreasing amount of β-FeOOH-like species. Influence of reducibility as well as surface and porosity characteristics seemed to have a more severe impact on catalytic performance for the nitrate samples. Lowest apparent activation energy in reduction for highest loaded nitrate sample correlated with highest acrolein selectivity, highest contribution of micropores to entire surface area, and roughest SBA-15 surface. Influence of amount of β-FeOOH-like species appeared to be inferior for the nitrate samples.

5.6 Interpretation of differences in structural and solid-state kinetic properties of Fe_xO_y/SBA-15 catalysts and their impact on catalytic performance

Figure 5.67 shows a schematic representation of proposed structure of iron oxidic species supported on SBA-15 dependent on iron loading and precursor. Based on this schematic representation, differences in reducibility and catalytic performance of Fe_xO_y/SBA-15 catalysts might be interpreted. In previous chapters, the degree of micropore filling was a measure for the strengths of interactions between iron oxidic species and SBA-15. These interactions between Fe_xO_y species and SBA-15 can be associated with the amount of Fe-O-Si bonds (Figure 5.67, (D)) and hydrogen bonds. Hydrogen bonds can be formed between either Fe-O structure units and surface hydroxyl groups (Si-OH) of SBA-15, or Fe-OH structure units and $[O-Si]_x$ structural motifs of SBA-15. Accordingly, stronger interactions between Fe_xO_y species and SBA-15 indicated a higher amount of both Fe-O-Si bonds and hydrogen bonds. The higher degree of micropore filling of citrate compared to corresponding nitrate samples with similar iron loading correlated with a higher amount of hydrogen bonds between iron oxidic species and SBA-15 due to the higher amount of β-FeOOH-like species (Figure 5.67, (A)).

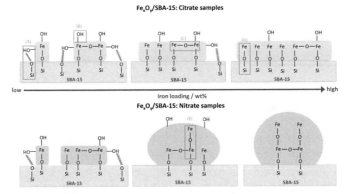

Figure 5.67: Schematic representation of structural differences of iron oxidic species supported on SBA-15 dependent on iron loading and precursor. Only the main structural motifs of the Fe_xO_y/SBA-15 catalysts are depicted.

As explained in chapter 2.2, selective oxidation of propene is assumed to proceed according to the so-called Mars-van-Krevelen mechanism. Thereby, metal oxide catalysts are partially reduced. Re-oxidation of the metal center of the catalyst has to be performed by molecular gas-phase oxygen. Conversion of gas-phase oxygen O_2 to nucleophilic O^{2-} must proceed quickly, otherwise electrophilic oxygen (O_2^-, O^-, O_2) will be formed. [7, 27, 169] Accordingly, a high electron transfer efficiency is required to enable a fast conversion of gas-phase oxygen. Selective oxidation of propene is assumed to be a four-electron process. [170] Hence, at least dimeric Fe_xO_y species are presumably required for generating acrolein as selective oxidation product. Fe-O-Fe structure units with bridging oxygen atoms in Fe_xO_y oligomers might determine the selectivity towards desired reaction product during selective oxidation reactions. Presence of Fe-O-Fe structure units might induce a higher selectivity towards desired reaction product. This could be attributed to a high electron transfer efficiency, and consequently, to the presence of neighboring iron centers in Fe_xO_y oligomers. Accordingly, partially reduced iron species might be stabilized by neighboring iron centers which share bridging oxygen atoms. Metal-oxygen-metal bonds are considered being nucleophilic, and hence, lead to attacking propene at the methyl group yielding acrolein as selective oxidation product. Conversely, higher amounts of electrophilic oxygen species result in total oxidation products, as well as C_2 aldehydes and C_2 acids due to attacking propene at the double bond. [27, 50]

For citrate samples, increasing iron loading induced a decreasing amount of β-FeOOH-like species and an increasing iron oxidic species size. Increasing Fe_xO_y species size correlated with an increasing amount of Fe-O-Fe structure units (Figure 5.67, (C)) and an increasing amount of Fe-O-Si structure units (Figure 5.67, (D)). This might explain the higher degree of micropore filling for higher loaded citrate samples. In literature, Fe-O-Si bonds are described as hardly reducible in contrast to Fe-O-Fe and Fe-OH bonds. [50] This agreed well with the highest reducibility for lowest loaded citrate samples. Both low amount of Fe-O-Si bonds and high amount of Fe-OH structure units (Figure 5.67, (A), (B)) obviously induced the high reducibility of lowest loaded citrate samples.

Within the citrate samples, increasing iron loading correlated with a decreasing acrolein selectivity. Hence, the higher amount of Fe-O-Fe structure units of higher loaded citrate samples was obviously not determining the catalytic performance. Highest acrolein selectivity for lowest loaded citrate samples might be induced by a favored H abstraction of the methyl group of propene due to the high amount of nucleophilic hydroxyl groups of Fe-OH structure units. Furthermore, the high amount of β-FeOOH-like species, and therefore, of neighboring hydroxyl groups might facilitate the recombination of hydroxyl groups and abstracted H from propene to form water and a reduced iron center. However, the high dispersion of small and isolated Fe_xO_y species may induce a lower electron transfer efficiency during selective oxidation reaction, and hence, facilitate formation of electrophilic surface oxygen species. This might explain the selectivity towards acetaldehyde. For higher loaded citrate samples, increased number of Fe-O-Fe structure units in Fe_xO_y oligomers yielded an increased electron transfer efficiency during selective oxidation of propene. However, the lower amount of Fe-OH structure units due to the lower amount of β-FeOOH-like species might result in a minor recombination of hydroxyl groups and abstracted H from propene. Despite a high electron transfer efficiency in Fe_xO_y oligomers, formation of electrophilic surface oxygen species might be facilitated, which induced a favored electrophilic attack at the double bond of propene. Accordingly, for citrate samples, selectivity towards acrolein decreased and that towards total oxidation products increased with increasing iron loading.

For nitrate samples, increasing iron loading induced an increasing iron oxidic species size and a decreasing amount of β-FeOOH-like species, similar to the citrate samples. However, Fe_xO_y species of nitrate samples were significantly larger than those of citrate samples and the amount of β-FeOOH-like species was minor. Higher loaded nitrate samples consisted mainly of Fe_xO_y oligomers and small α-Fe_2O_3 particles. [FeO_6] units were not only edge-shared but also corner-shared (Figure 5.67, (E)), yielding a lower amount of Fe-O-Si bonds (Figure 5.67, (D)). Furthermore, the number of Fe-O-Fe structure units (Figure 5.67, (C)) was significantly increased compared to lower loaded nitrate samples and to citrate samples. Hence, reducibility of highest loaded nitrate samples was high and degree of micropore filling was low. Formation of nucleophilic surface O^{2-} species might be facilitated due to the high reducibility of higher aggregated Fe_xO_y species. Hence, nucleophilic attacking propene at the methyl group became more probable. This might explain the increasing acrolein selectivity with increasing iron loading within the nitrate samples. Conversely, lower iron loadings within the nitrate samples correlated with a higher amount of dimeric and smaller Fe_xO_y oligomers. Consequently, a higher number of Fe-O-Si bonds was formed, which induced a higher degree of micropore filling. The increased number of Fe-O-Si bonds together with the decreased number of Fe-O-Fe structure units resulted in both a lower reducibility and a lower acrolein selectivity compared to higher loaded nitrate samples. Moreover, lower reducibility and lower electron transfer efficiency for lower loaded nitrate samples might induce a facilitated formation of electrophilic surface oxygen species, which yielded a higher selectivity towards total oxidation products.

For Fe_xO_y/SBA-15 catalysts, both the amount of hydroxyl groups and that of Fe-O-Fe structure units was crucial for the differences in catalytic performance dependent on iron loading and precursor. The influence of β-FeOOH-like species was more important for both lower iron loadings and citrate samples. Conversely, reducibility became the dominating factor for catalytic performance of nitrate samples, possessing lower amounts of β-FeOOH-like species.

5.7 Conclusion

Synthesis of Fe_xO_y/SBA-15 catalysts using two different precursors, $(NH_4, Fe(III))$ citrate and Fe(III) nitrate nonahydrate, induced small and highly dispersed iron oxidic species located in the pore system of SBA-15. Structure of support material remained unaffected during synthesis. Detailed structural characterization of Fe_xO_y/SBA-15 catalysts revealed precursor-dependent and iron loading-dependent differences in surface and porosity characteristics, local structure, and types of iron oxidic species. In addition to iron loading-dependent particle size effect, decreasing iron loading correlated with an increasing amount of β-FeOOH-like species independent of the precursor. Predominantly octahedrally coordinated Fe^{3+} species in small dimeric or oligomeric Fe_xO_y with a local geometry similar to that of α-Fe_2O_3 or β-FeOOH were identified as major structural motif in all Fe_xO_y/SBA-15 catalysts.

The citrate precursor induced smaller iron oxidic species, a higher degree of micropore filling, and a smoother SBA-15 surface compared to the nitrate precursor. Precursor-dependent differences in surface and porosity properties were ascribed to different strengths of chelating effect of the two precursors. More pronounced chelating effect of the citrate precursor induced stronger interactions between precursor molecules and support material. This yielded a favored formation of hydrogen bonds between citrate precursor and superficial silanol groups of SBA-15. Consequently, the citrate precursor induced higher amounts of β-FeOOH-like species due to the chemical memory effect of Fe_xO_y/SBA-15.

For Fe_xO_y/SBA-15 catalysts obtained from citrate precursor, amount of β-FeOOH-like species directly correlated with catalytic performance. Onset of acrolein formation under propene oxidation conditions was associated with amount of β-FeOOH-like species exceeding 50%. Within the citrate samples, increasing acrolein selectivity correlated with increasing amount of β-FeOOH-like species and decreasing iron loading. Higher amounts of nucleophilic hydroxyl groups of Fe-OH structure units might induce a favored H abstraction of the methyl group of propene. Moreover, high amounts of neighboring hydroxyl groups, due to the high amount of β-FeOOH-like species, might facilitate the recombination of hydroxyl groups and abstracted H from propene to form water and a reduced iron center. Reaction rate of propene conversion for citrate samples was higher compared to nitrate samples with similar iron loading. This agreed with precursor-dependent particle size effect, and further with precursor-dependent differences in types of iron oxidic species. For selectivity towards acrolein, a combination of several effects was considered responsible for precursor-dependent differences. For elucidating the influence of precursor-dependent effects on catalytic performance, solid-state kinetic analysis was particularly helpful.

Applicability of solid-state kinetic analysis to supported iron oxidic species was shown and results agreed well with those from structural characterization of Fe_xO_y/SBA-15 catalysts. Additionally, solid-state kinetic properties of reduction of Fe_xO_y/SBA-15 catalysts enabled a further insight into precursor-dependent differences. For Fe_xO_y/SBA-15 catalysts obtained from nitrate precursor, an increasing iron loading correlated with a better reducibility and a more complex reduction mechanism. Larger iron oxidic species resulting for higher iron loadings within the nitrate samples induced a lower degree of micropore filling together with a rougher SBA-15 surface. This was indicative of weaker interactions between iron oxidic species and support material. These weaker interactions correlated with decreasing apparent activation energy in reduction with increasing iron loading for Fe_xO_y/SBA-15 catalysts obtained from nitrate precursor. Conversely, within the citrate samples, decreasing apparent activation energy in reduction correlated with decreasing iron loading, and hence, with smaller iron oxidic species and higher amounts of β-FeOOH-like species. For iron loadings higher than 2.5wt%, the citrate precursor induced higher values of apparent activation energy in reduction compared to the nitrate precursor. The better reducibility of the nitrate samples was furthermore reflected in higher amounts of total oxidation products under catalytic conditions. Nevertheless, for nitrate samples with significantly lower amounts of β-FeOOH-like species, reducibility was crucial for acrolein selectivity. Within the nitrate samples, increasing reducibility, together with decreasing degree of micropore filling and increasing SBA-15 surface roughness induced an increasing acrolein selectivity. Higher acrolein selectivity for higher loaded nitrate samples might be ascribed to a facilitated formation of nucleophilic O^{2-} species, and consequently, more probable nucleophilic attacking propene at the methyl group. Detailed characterization of Fe_xO_y/SBA-15 catalysts under reaction conditions suggested not only one single effect, but a combination of particle size effect, interactions between iron oxidic species and SBA-15, amount of β-FeOOH-like species, and reducibility influencing catalytic performance. For iron loadings up to 7.2wt%, using the citrate precursor led to an increased selectivity towards desired partial oxidation product acrolein. Optimal surface coverage amounted to 0.47 Fe atoms per nm^2. Conversely, for iron loadings higher than 7.2wt%, using the nitrate precursor induced a favored selective oxidation with an optimal surface coverage of 1.58 Fe atoms per nm^2.

Variations in local structure of Fe_xO_y/SBA-15 catalysts after catalytic reaction were predominantly ascribed to variations in phase composition. Fe_xO_y/SBA-15 catalysts after selective oxidation of propene still consisted of α-Fe_2O_3-like and β-FeOOH-like species, but the amount of β-FeOOH-like species was significantly increased. Besides varied phase composition after catalytic reaction, a slightly increased degree of aggregation of iron oxidic species seemed likely, especially for citrate samples.

In situ characterization showed Fe_xO_y/SBA-15 being suitable model catalysts for elucidating correlations between structural characteristics, reducibility, catalytic performance in selective oxidation of propene, and additionally, solid-state kinetic properties. Not only varying iron loading but also varying Fe(III) precursor during synthesis of Fe_xO_y/SBA-15 catalysts had a significant influence on their structural characteristics, reducibility, and catalytic performance in the selective oxidation of propene.

6 Influence of powder layer thickness during calcination on Fe$_x$O$_y$/SBA-15 catalysts

6.1 Introduction

In literature, investigations of iron-containing catalysts were reported regarding the correlation between iron species size, precursor and promotor effect, as well as calcination conditions and their catalytic performance. [25, 42, 43, 45, 50, 56] For copper oxides supported on SBA-15, an influence of powder layer thickness during calcination has been revealed by Koch et al. [110]. They reported a distinct effect of calcination in a thick powder layer on size and electronic structure of supported copper oxides. Thick layer calcination was associated with an increased copper oxide particle size compared to thin layer calcination and a decreased reducibility. Detailed characterization of Fe$_x$O$_y$/SBA-15 catalysts in the previous chapter revealed a distinct influence of the precursor on structural and catalytic, as well as solid-state kinetic properties of supported iron oxidic species. In this chapter, influence of powder layer thickness during calcination as further synthesis parameter was investigated. Therefore, structural characterization was conducted by combination of various characterization methods, such as N$_2$ physisorption, XRD, and DR-UV-Vis spectroscopy. Additionally, solid-state kinetic analysis was applied to elucidate correlations between powder layer thickness during calcination, structure, and reducibility of supported iron oxidic species. These correlations were deduced with respect to precursor and iron loading. Moreover, Fe$_x$O$_y$/SBA-15 catalysts were characterized under propene oxidation conditions for investigating the influence of powder layer thickness during calcination on catalytic performance.

In this chapter, it will be shown that varying powder layer thickness during calcination affects iron oxidic species, surface and porosity characteristics, and furthermore, reducibility and catalytic performance of Fe$_x$O$_y$/SBA-15 catalysts. It will be clarified that the influence of powder layer thickness during calcination on Fe$_x$O$_y$/SBA-15 catalysts depends on both Fe(III) precursor and iron loading. Additionally, solid-state kinetic analysis of the reduction of Fe$_x$O$_y$/SBA-15 catalysts will further corroborate the results from structural characterization. Correlation between varying powder layer thickness during calcination, structure, reducibility, and catalytic performance of Fe$_x$O$_y$/SBA-15 catalysts will be highlighted.

6.2 Experimental

6.2.1 Sample preparation

Mesoporous silica SBA-15 was prepared according to Zhao et al. [33, 34] as described in chapter 4.2.1. Synthesis of supported iron oxidic species on SBA-15 using two different precursors was performed similar to synthesis procedure explained in chapter 5.2.1. However, after drying in air for 24 h, calcination was carried out at 723 K for 2 h in either thin powder layer or thick powder layer. In this work, regular calcination mode of Fe$_x$O$_y$/SBA-15 catalysts was thin powder layer calcination, whereby powder layer thickness was 0.3 cm. Conversely, for thick powder layer calcination, powder layer thickness was 1.3 cm. Samples calcined in thick powder layer were denoted with the ending "_Th" to clarify the different layer thickness during calcination compared to corresponding Fe_xO_y/SBA-15 samples. Denotation of Fe_xO_y/SBA-15 as citrate or nitrate samples dependent on used precursor was retained from previous chapter. Accordingly, Fe_xO_y/SBA-15 samples calcined in thick powder layer obtained from citrate precursor were denoted citrate_Th samples and those obtained from nitrate precursor nitrate_Th samples.

After synthesis of Fe_xO_y/SBA-15_Th samples, iron loadings were confirmed by X-ray fluorescence spectroscopy. Furthermore, CHN analysis was performed to confirm a complete decomposition of the used precursor. Citrate samples calcined in thin powder layer were light orange brownish, whereby color intensity increased with increasing iron loading. Corresponding citrate samples calcined in thick powder layer showed no significant color change, except for highest loaded citrate sample. 10.7wt% Fe_Citrate_Th was reddish orange and color intensity was increased after thick powder layer calcination. Conversely, using the nitrate precursor resulted in a more pronounced color change dependent on layer thickness during calcination. Color of nitrate samples calcined in thin powder layer was orange for the lowest loaded sample, reddish orange for 7.2wt% Fe_Nitrate, and dark red for 9.3wt% Fe_Nitrate. For nitrate samples, calcination in thick powder layer induced an increased color intensity independent of the iron loading and a change in color to more reddish for an iron loading of 2.0wt%. Nitrate_Th samples with iron loadings higher than 2.0wt% appeared brick-red, whereby color intensity increased with increasing iron loading. Figure 6.1 illustrates the change in color of Fe_xO_y/SBA-15 samples for thin and thick powder layer calcination dependent on both precursor and iron loading.

	Citrate samples		Nitrate samples		
wt%	thin	thick	thick	thin	wt%

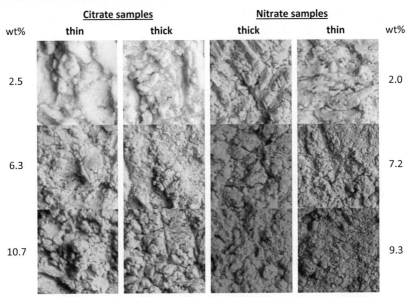

Figure 6.1: Fe$_x$O$_y$/SBA-15 samples calcined in thin or thick powder layer using the citrate and nitrate precursor, respectively. Increasing iron loading (wt%) from top to bottom.

6.2.2 Sample characterization

X-ray fluorescence analysis

Quantitative analysis of the metal oxide loadings on SBA-15 was conducted by X-ray fluorescence spectroscopy on an AXIOS X-ray spectrometer (2.4 kW model, PANalytical), equipped with a Rh K$_\alpha$ X-ray source, a gas flow detector, and a scintillation detector. Prior to measurements, samples were mixed with wax (Hoechst wax C micropowder, Merck), ratio 1:1, and pressed into pellets of 13 mm diameter. Quantification was performed by standardless analysis using the software package SuperQ5 (PANalytical).

CHN elemental analysis

Elemental contents of C, H, and N were determined using a Thermo FlashEA 1112 Organic Elemental Analyzer (ThermoFisher Scientific) with CHNS-O configuration. Measurements were conducted in collaboration with the measuring center at the institute of chemistry at TU Berlin.

Nitrogen physisorption

Nitrogen adsorption/desorption isotherms were measured at 77 K using a BELSORP Mini II (BEL Inc. Japan). Prior to measurements, the samples were pre-treated under reduced pressure (10^{-2} kPa) at 368 K for 35 min and kept under the same pressure at 448 K for 15 h (BELPREP II vac).

Transmission electron microscopy

Transmission electron microscopy (TEM) images were recorded on a FEI Tecnai G^2 20 S-TWIN microscope equipped with a LaB_6 cathode and a 1 k x 1 k CCD camera (GATAN MS794). Acceleration voltage was set to 220 kV and samples were prepared on 300 mesh Cu grids with Holey carbon film. Measurements were conducted in collaboration with ZELMI (Zentraleinrichtung für Elektronenmikroskopie) at TU Berlin.

Powder X-ray diffraction

Powder X-ray diffraction patterns were obtained using an X'Pert PRO diffractometer (PANalytical, 40 kV, 40 mA) in theta/theta geometry equipped with a solid-state multi-channel detector (PIXel). Cu $K_α$ radiation was used. Wide-angle diffraction scans were conducted in reflection mode. Small-angle diffraction patterns were measured in transmission mode from 0.4° through 6° 2θ in steps of 0.013° 2θ with a sampling time of 90 s/step.

Diffuse reflectance UV-Vis spectroscopy

Diffuse reflectance UV-Vis (DR-UV-Vis) spectroscopy was conducted on a two-beam spectrometer (V-670, Jasco) using a barium sulfate coated integration sphere. (Scan speed 100 nm/min, slit width 5.0 nm (UV-Vis) and 20.0 nm (NIR), and spectral region 220-2000 nm). SBA-15 was used as white standard for all samples.

Temperature-programmed reduction

Temperature-programmed reduction (TPR) was performed using a BELCAT_B (BEL Inc. Japan). Samples were placed on silica wool in a silica glass tube reactor. Evolving water was trapped using a molecular sieve (4 Å). Gas mixture consisted of 5% H_2 in 95% Ar with a total gas flow of 40 ml/min. Heating rates used were 5, 10, 15, and 20 K/min to 1223 K. A constant initial sample weight of 0.03 g was used and H_2 consumption was continuously monitored by a thermal conductivity detector.

Catalytic characterization

Quantification of catalytic activity in selective oxidation of propene was performed using a laboratory fixed-bed reactor connected to an online gas chromatography system (CP-3800, Varian) and a mass spectrometer (Omnistar, Pfeiffer Vacuum). The fixed-bed reactor consisted of a SiO_2 tube (30 cm length, 9 mm inner diameter) placed vertically in a tube furnace. A P3 frit was centered in the SiO_2 tube in the isothermal zone, where the sample was placed. The samples were diluted with boron nitride (99.5%, Alfa Aesar) to achieve a constant volume in the reactor and to minimize thermal effects. Overall sample masses were 0.25 g. The reactor was operated at low propene conversion levels (5-10%) to ensure differential reaction conditions. For catalytic testing in selective oxidation of propene, a gas mixture of 5% propene and 5% oxygen in helium was used in a temperature range of 298-653 K with a heating rate of 5 K/min. Reactant gas flow rates of propene, oxygen, and helium were adjusted through separate mass flow controllers (Bronkhorst) to a total flow of 40 ml/min. All gas lines and valves were preheated to 473 K. Hydrocarbons and oxygenated reaction products were analyzed using a Carbowax 52CB capillary column, connected to an Al_2O_3/MAPD capillary column (25 m x 0.32 mm) or a fused silica restriction (25 m x 0.32 mm), each connected to a flame ionization detector (FID). Permanent gases (CO, CO_2, N_2, O_2) were separated and analyzed using a "Permanent Gas Analyzer" (CP-3800, Varian) with a Hayesep Q (2 m x 1/8") and a Hayesep T packed column (0.5 m x 1/8") as precolumns combined with a back flush. For separation, a Hayesep Q packed column (0.5 m x 1/8") was connected via a molecular sieve (1.5 m x 1/8") to a thermal conductivity detector (TCD). Additionally, product and reactant gas flow were continuously monitored by a connected mass spectrometer (Omnistar, Pfeiffer) in a multiple ion detection mode. Details on calculation of catalytic parameters are described in chapter 5.2.2.

6.3 Structural characterization of Fe_xO_y/SBA-15_Th catalysts

6.3.1 Pore structure and surface properties

Iron oxide dispersion

Nitrogen physisorption measurements were conducted for characterizing Fe_xO_y/SBA-15_Th samples regarding their surface properties and pore structure. Nitrogen adsorption/desorption isotherms of Fe_xO_y/SBA-15_Th samples dependent on used precursor are depicted in Figure 6.2. Similar to SBA-15 support material and Fe_xO_y/SBA-15 samples (see chapter 4.3 and 5.3.1), Fe_xO_y/SBA-15_Th samples exhibited type IV nitrogen adsorption/desorption isotherms indicating mesoporous materials. [68] Adsorption and desorption branches were nearly parallel at the hysteresis loop, as expected for regularly shaped pores. Hysteresis loops of Fe_xO_y/SBA-15_Th samples were identified being of type H1 according to IUPAC. [68]

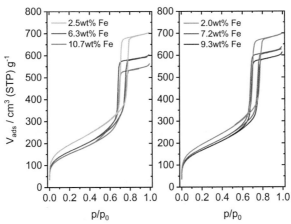

Figure 6.2: N_2 adsorption/desorption isotherms of Fe_xO_y/SBA-15_Th samples. Left: Citrate_Th samples. Right: Nitrate_Th samples.

BET [69] and BJH [70] methods were applied to nitrogen physisorption data for determining specific surface area and pore radius distribution, respectively. Fe_xO_y/SBA-15_Th samples, as well as Fe_xO_y/SBA-15 samples and SBA-15, exhibited high specific surface areas, $a_{s,BET}$, with narrow pore size distributions. Independent of the precursor, lowest loaded Fe_xO_y/SBA-15_Th samples possessed highest specific surface area and increasing the iron loading induced a decrease in specific surface area. Compared to Fe_xO_y/SBA-15 samples, specific surface area of Fe_xO_y/SBA-15_Th samples further decreased (Table 6.1). The decrease in specific surface area induced by thick layer calcination was more pronounced for the citrate samples.

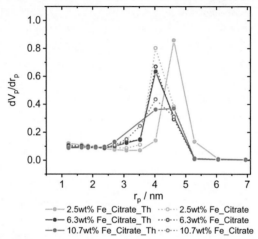

Figure 6.3: Pore radius distribution of citrate_Th and corresponding citrate samples calculated by BJH method.

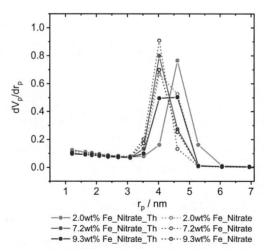

Figure 6.4: Pore radius distribution of nitrate_Th and corresponding nitrate samples calculated by BJH method.

Pore radius distributions for citrate_Th and nitrate_Th samples compared to corresponding Fe_xO_y/SBA-15 samples are depicted in Figure 6.3 and Figure 6.4. A narrow pore radius distribution was observed for all Fe_xO_y/SBA-15_Th and Fe_xO_y/SBA-15 samples. Thick layer calcination induced a shift of the main maximum in pore radius distribution to larger pore radius, except for samples 6.3wt% Fe_Citrate and 7.2wt% Fe_Nitrate. For these two samples, pore radius distribution was unaffected by layer thickness during calcination. The most pronounced influence of layer thickness during calcination was observed for the lowest loaded citrate and nitrate sample. Narrow pore radius distribution of these samples was shifted to larger pore radius after thick layer calcination. Conversely, for the highest loaded citrate and nitrate sample, pore radius distribution after thick layer calcination appeared slightly broadened and slightly shifted to larger pore radius. Additionally, pore volume, V_{pore}, was determined and summarized in Table 6.1. Thick layer calcination induced a decrease in pore volume compared to thin layer calcination. Conversely, 7.2wt% Fe_Nitrate_Th and 9.3wt% Fe_Nitrate_Th showed a slight increase in V_{pore}. For Fe_xO_y/SBA-15_Th samples, the decrease in specific surface area together with the decrease in pore volume with increasing iron loading indicated the presence of iron oxidic species in the mesopores of SBA-15. Despite slightly increased V_{pore}, indicating some mesopores might remain unfilled, iron oxidic species being mainly located in the mesopores of SBA-15 seemed also likely for 7.2wt% Fe_Nitrate_Th and 9.3wt% Fe_Nitrate_Th. Location of iron oxidic species in the pore system of SBA-15 was furthermore confirmed by transmission electron microscopy measurements of Fe_xO_y/SBA-15_Th samples. TEM micrograph of 10.7wt% Fe_Citrate_Th is depicted in Figure 6.5, left. Absence of detectable iron oxide particles in TEM micrographs of all citrate_Th samples confirmed successful synthesis of small iron oxidic species in the pore system of SBA-15 with no iron oxidic species on the external surface of SBA-15. Sample 2.0wt% Fe_Nitrate_Th showed a similar TEM micrograph to citrate_Th samples. Conversely, for higher loaded nitrate_Th samples, dark contrasts were discernible in TEM micrographs as depicted for 9.3wt% Fe_Nitrate_Th in Figure 6.5, right. These dark contrasts were indicative of higher aggregated iron oxidic species and agreed well with the slight increase in V_{pore}. Moreover, TEM micrographs of all Fe_xO_y/SBA-15_Th samples showed characteristic hexagonal ordering of mesopores in the two-dimensional framework of SBA-15. Results from TEM measurements agreed with those from N_2 physisorption and confirmed the SBA-15 structure to be preserved after thick layer calcination.

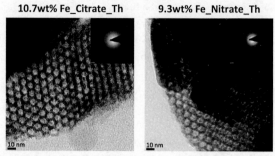

Figure 6.5: TEM micrographs of 10.7wt% Fe_Citrate_Th (left) and 9.3wt% Fe_Nitrate_Th (right). Inset depicts small-angle electron diffraction (SAED) image.

Iron oxide surface coverage of Fe_xO_y/SBA-15_Th samples is summarized in Table 6.1. A slightly increased surface coverage of Fe_xO_y/SBA-15_Th compared to corresponding Fe_xO_y/SBA-15 samples was observed. This was indicative of a slightly increased aggregation of iron oxidic species on SBA-15 induced by thick layer calcination. Nevertheless, surface coverage of all Fe_xO_y/SBA-15_Th samples was comparatively low. This corroborated the results from TEM and N_2 physisorption measurements and excluded a monolayer coverage.

Table 6.1: Surface and porosity characteristics of Fe_xO_y/SBA-15_Th samples. Specific surface area, $a_{s,BET}$, pore volume, V_{pore}, and surface coverage, $\Phi_{Fe\ atoms}$. Differences between Fe_xO_y/SBA-15_Th and corresponding Fe_xO_y/SBA-15 samples are denoted as Δ values.

	$a_{s,BET}$ / $m^2\,g^{-1}$	$\Delta a_{s,BET}$ / $m^2\,g^{-1}$	V_{pore} / $cm^3\,g^{-1}$	ΔV_{pore} / $cm^3\,g^{-1}$	$\Phi_{Fe\ atoms}$ / Fe atoms nm^{-2}
2.5wt% Fe_Citrate_Th	700.0 ± 0.7	-25.5	1.087 ± 0.001	-0.025	0.38
6.3wt% Fe_Citrate_Th	592.6 ± 0.6	-58.3	0.933 ± 0.001	-0.035	1.14
10.7wt% Fe_Citrate_Th	550.2 ± 0.6	-55.5	0.865 ± 0.001	-0.033	2.09
2.0wt% Fe_Nitrate_Th	701.7 ± 0.7	-1.3	1.079 ± 0.001	-0.017	0.30
7.2wt% Fe_Nitrate_Th	633.4 ± 0.6	-14.4	0.987 ± 0.001	0.016	1.22
9.3wt% Fe_Nitrate_Th	606.0 ± 0.7	-27.1	0.943 ± 0.001	0.004	1.65

Contribution of micropores to entire surface area

Ratio of mesopore surface area, a_{pore}, calculated by BJH method [70] and specific surface area, $a_{s,BET}$, calculated by BET method [69] of Fe_xO_y/SBA-15_Th samples was used for estimating the contribution of micropores to the entire surface area of SBA-15 (Table 6.2). [71, 110] Furthermore, the effect of layer thickness during calcination on ratio of $a_{pore}/a_{s,BET}$ was determined by calculating the difference between Fe_xO_y/SBA-15_Th and Fe_xO_y/SBA-15 samples. Thick layer calcination induced an increased ratio of $a_{pore}/a_{s,BET}$, except for 2.0wt% Fe_Nitrate_Th and 2.5wt% Fe_Citrate_Th. These lowest loaded Fe_xO_y/SBA-15_Th samples possessed the same ratio of $a_{pore}/a_{s,BET}$, which was slightly decreased compared to that observed for corresponding Fe_xO_y/SBA-15 samples. Accordingly, thick layer calcination of lowest loaded samples correlated with an increased contribution of micropores to entire surface area, and hence, a lower degree of micropore filling. For iron loadings higher than 2.5wt%, citrate_Th samples showed higher ratios of $a_{pore}/a_{s,BET}$ compared to nitrate_Th samples with similar iron loading. Apparently, using the citrate precursor resulted in stronger interactions between SBA-15 and Fe(III) atoms of the precursor, and hence, in higher amounts of micropore filling, independent of layer thickness during calcination. This was ascribed to the more pronounced chelating effect the citrate precursor. Moreover, $Fe_xO_y/SBA-15\_Th$ samples showed a precursor-dependent trend in ratio of $a_{pore}/a_{s,BET}$ as function of iron loading. For citrate_Th samples, increasing the iron loading correlated with an increasing ratio of $a_{pore}/a_{s,BET}$. Conversely, for nitrate_Th samples, ratio of $a_{pore}/a_{s,BET}$ only increased up to 7.2wt% iron, but decreased with further increasing iron loading. Influence of thick layer calcination on contribution of micropores to entire surface area was more pronounced for the citrate precursor.

Table 6.2: Contribution of micropores to entire surface area and surface roughness for $Fe_xO_y/SBA-15\_Th$. Ratio of mesopore surface area, a_{pore}, and specific surface area, $a_{s,BET}$, as measure of micropore contribution to the entire surface of SBA-15, differences in fractal dimension, ΔD_f, between SBA-15 and corresponding $Fe_xO_y/SBA-15\_Th$ samples as measure of the roughness of the surface. Differences between $Fe_xO_y/SBA-15\_Th$ and corresponding $Fe_xO_y/SBA-15$ samples are denoted as Δ values.

	$a_{pore}/a_{s,BET}$	$\Delta(a_{pore}/a_{s,BET})$	ΔD_f	$\Delta(\Delta D_f)$
2.5wt% Fe_Citrate_Th	0.86	-0.02	-0.02 ± 0.02	-0.14
6.3wt% Fe_Citrate_Th	0.94	0.05	0.32 ± 0.05	0.11
10.7wt% Fe_Citrate_Th	0.95	0.04	0.41 ± 0.08	0.26
2.0wt% Fe_Nitrate_Th	0.86	-0.02	0.02 ± 0.02	-0.04
7.2wt% Fe_Nitrate_Th	0.90	0.02	0.20 ± 0.01	0.13
9.3wt% Fe_Nitrate_Th	0.88	0.02	0.10 ± 0.02	0.16

Surface roughness

Modified FHH method was additionally applied for analyzing nitrogen physisorption data of $Fe_xO_y/SBA-15\_Th$ samples. Resulting fractal dimension D_f was used as a measure of the roughness of the surface. [71, 72] As for $Fe_xO_y/SBA-15$ samples and SBA-15, the fractal dimension of $Fe_xO_y/SBA-15\_Th$ samples ranged between 2 and 3 indicating a rough surface. ΔD_f values were calculated as difference between D_f values of SBA-15 and those of $Fe_xO_y/SBA-15\_Th$ samples for determining the influence of supported iron oxidic species on surface roughness of the support material. Additionally, differences between ΔD_f values of $Fe_xO_y/SBA-15\_Th$ and corresponding $Fe_xO_y/SBA-15$ samples were calculated for elucidating the effect of layer thickness during calcination on surface roughness of support material (Table 6.2). For lowest loaded samples, 2.5wt% Fe_Citrate_Th and 2.0wt% Fe_Nitrate_Th, ΔD_f decreased indicating an increased SBA-15 surface roughness. This increased surface roughness complied with the decreased ratio of $a_{pore}/a_{s,BET}$. For iron loadings higher than 2.5wt%, thick layer calcination induced significantly higher ΔD_f values compared to thin layer calcination, and hence, a smoother SBA-15 surface. A smoothing of SBA-15 surface for $Fe_xO_y/SBA-15\_Th$ samples agreed with an increased degree of micropore filling. This was induced by higher aggregated and less dispersed iron oxidic species of Fe_xO_y/SBA-15_Th samples. Moreover, compared to the nitrate_Th samples, SBA-15 surface of the citrate_Th samples was smoother. Whereas ΔD_f of the citrate_Th samples increased with increasing iron loading, those of the nitrate_Th samples increased up to 7.2wt% iron and subsequently decreased for highest loaded nitrate_Th sample (Figure 6.6). This decrease in ΔD_f for 9.3wt% Fe_Nitrate_Th might result from a significantly increased iron oxidic species size implied by an increased V_{pore}.

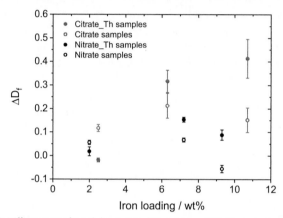

Figure 6.6: Differences in fractal dimension, ΔD_f, between SBA-15 support material and corresponding Fe_xO_y/SBA-15_Th and Fe_xO_y/SBA-15 samples, dependent on precursor and iron loading.

Analogous to Fe_xO_y/SBA-15 samples, precursor-dependent differences in surface roughness of the support material of Fe_xO_y/SBA-15_Th samples may be explained by different strengths of chelating effect of the two Fe(III) precursors (see chapter 5.3.1). However, thick layer calcination enhanced the aggregation and reduced the dispersion of iron oxidic species.

6.3.2 Long-range ordered structure of Fe_xO_y/SBA-15_Th

Long-range ordered structure of Fe_xO_y/SBA-15_Th samples was investigated by wide-angle and small-angle powder X-ray diffraction. Figure 6.7 depicts wide-angle X-ray diffraction patterns of Fe_xO_y/SBA-15_Th samples. For all Fe_xO_y/SBA-15_Th as well as for all Fe_xO_y/SBA-15 samples, a broad diffraction peak at 23° 2θ was observed, corresponding to the amorphous silica structure of SBA-15. [49] XRD patterns of all citrate samples, originating from either thin layer calcination or thick layer calcination, showed no further diffraction peaks (Figure 5.8 and Figure 6.7). This was indicative of only small and highly dispersed iron oxidic species being present in citrate samples independent of layer thickness during calcination. XRD patterns of Fe_xO_y/SBA-15 samples originating from nitrate samples calcined in thin layer also showed no further diffraction peaks indicative of no long-range ordered phases (Figure 5.8). Conversely, XRD pattern of 9.3wt% Fe_Nitrate_Th showed small diffraction peaks of crystalline α-Fe_2O_3

(Figure 6.7). Furthermore, for 7.2wt% Fe_Nitrate_Th, two main diffraction peaks of α-Fe$_2$O$_3$ were discernible, while 2.0wt% Fe_Nitrate_Th showed no further diffraction peaks. Accordingly, using the nitrate precursor in thick layer calcination for iron loadings higher than 2.0wt% induced significantly larger and less dispersed iron oxidic species on SBA-15 compared to those obtained from citrate precursor.

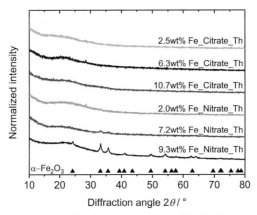

Figure 6.7: Wide-angle X-ray diffraction patterns of Fe$_x$O$_y$/SBA-15_Th samples. Diffraction peaks corresponding to α-Fe$_2$O$_3$ are depicted as black triangles.

Small-angle X-ray diffraction patterns of Fe$_x$O$_y$/SBA-15_Th samples exhibited three characteristic diffraction peaks, (10l), (11l), and (20l), corresponding to the two-dimensional hexagonal symmetry of SBA-15 (Figure 6.8). [32, 33, 101] Hence, SBA-15 structure was retained during synthesis with calcination in thick powder layer. Lattice constant, a_0, of the hexagonal unit cell was determined from the diffraction peaks (10l), (11l), and (20l), using equation (3.10) for all Fe$_x$O$_y$/SBA-15_Th samples (Table 6.3). As observed for Fe$_x$O$_y$/SBA-15 samples (see chapter 5.3.2), for Fe$_x$O$_y$/SBA-15_Th samples, using the nitrate precursor resulted in slightly higher values of lattice constant and wall thickness, d_w, compared to the citrate precursor. Thick layer calcination induced slightly decreased values of lattice constant and wall thickness compared to those of thin layer calcination. Nevertheless, values of a_0 and d_w of Fe$_x$O$_y$/SBA-15_Th samples were of comparable order of magnitude to those of SBA-15, indicating preserved pore structure of SBA-15 after synthesis.

Table 6.3: Lattice constant, a_0, and wall thickness, d_w, of Fe_xO_y/SBA-15_Th samples. Differences between values of Fe_xO_y/SBA-15_Th and those of corresponding Fe_xO_y/SBA-15 samples are denoted as Δ values.

	a_0 / nm	Δa_0 / nm	d_w / nm	Δd_w / nm
2.5wt% Fe_Citrate_Th	11.04 ± 0.01	-0.11	1.82 ± 0.01	-1.27
6.3wt% Fe_Citrate_Th	10.96 ± 0.02	-0.06	2.90 ± 0.02	-0.06
10.7wt% Fe_Citrate_Th	10.92 ± 0.02	-0.02	1.80 ± 0.02	-1.09
2.0wt% Fe_Nitrate_Th	11.12 ± 0.01	-0.04	1.90 ± 0.01	-1.20
7.2wt% Fe_Nitrate_Th	11.10 ± 0.01	-0.01	3.04 ± 0.01	-0.01
9.3wt% Fe_Nitrate_Th	11.04 ± 0.01	-0.06	1.82 ± 0.01	-1.21

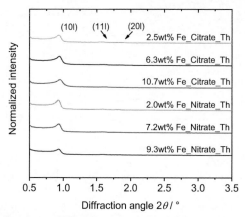

Figure 6.8: Small-angle X-ray diffraction patterns of Fe_xO_y/SBA-15_Th samples.

6.3.3 Diffuse reflectance UV-Vis spectroscopy

DR-UV-Vis spectra of Fe_xO_y/SBA-15_Th and corresponding Fe_xO_y/SBA-15 samples are depicted in Figure 6.9. All Fe_xO_y/SBA-15_Th samples showed a broad absorption resulting from overlapping absorption bands. Analogous to Fe_xO_y/SBA-15 samples (see chapter 5.3.3), this intense and broad absorption of all Fe_xO_y/SBA-15_Th samples was ascribed to charge transfer (CT) transitions between oxygen ligands (O^{2-}) and central iron atoms (Fe^{3+}) in iron oxidic species on SBA-15. Using the citrate precursor yielded a slight red-shift of the absorption in DR-UV-Vis spectra of Fe_xO_y/SBA-15_Th compared to corresponding Fe_xO_y/SBA-15 samples (Figure 6.9). Conversely, using the nitrate precursor affected more pronounced differences in DR-UV-Vis spectra dependent on layer thickness during calcination. Nitrate_Th samples showed a distinct red-shift of the absorption in DR-UV-Vis spectra compared to corresponding nitrate samples. For 9.3wt% Fe_Nitrate_Th, a weak absorption band at low wavenumbers (< 20000 cm^{-1}), similar to that in DR-UV-Vis spectrum of the mechanical mixture (Figure 5.14), was observed. This absorption band corresponded to symmetry and spin forbidden single electron d-d transitions in Fe^{3+}. [118, 119] Hence, for this highest loaded nitrate_Th sample, at least small amounts of crystalline α-Fe_2O_3 were present, which was confirmed by diffraction peaks of crystalline α-Fe_2O_3 (Figure 6.7). Differences in DR-UV-Vis spectra of Fe_xO_y/SBA-15_Th and corresponding Fe_xO_y/SBA-15 samples indicated a precursor-dependent influence of layer thickness during calcination on resulting iron oxidic species on SBA-15. Hence, results from DR-UV-Vis spectroscopy corroborated those from N_2 physisorption, TEM, and XRD measurements.

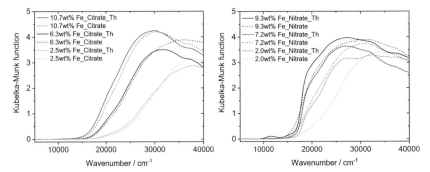

Figure 6.9: DR-UV-Vis spectra of Fe_xO_y/SBA-15_Th and corresponding Fe_xO_y/SBA-15 samples at various iron loadings. Left: Citrate and citrate_Th samples. Right: Nitrate and nitrate_Th samples.

As explained in chapter 5.3.3, values of the DR-UV-Vis edge energy can be used for estimating metal oxide particle size. Therefore, edge energy values of Fe_xO_y/SBA-15_Th samples were determined from DR-UV-Vis spectra (Figure 6.9) according to Weber [114] DR-UV-Vis edge energies for Fe_xO_y/SBA-15_Th samples, together with those of corresponding Fe_xO_y/SBA-15 samples and that of the mechanical mixture, Fe_2O_3/SBA-15, are depicted in Figure 6.10. Fe_xO_y/SBA-15_Th samples showed significantly lower edge energies compared to Fe_xO_y/SBA-15 samples, indicative of a particle size effect dependent on layer thickness during calcination. Comparing differences in edge energies between Fe_xO_y/SBA-15_Th and Fe_xO_y/SBA-15 samples revealed a more pronounced influence of layer thickness during calcination for the nitrate precursor. Additionally, decreasing iron loading correlated with an increasing difference in edge energy values between Fe_xO_y/SBA-15_Th and corresponding Fe_xO_y/SBA-15 sample. Hence, thick layer calcination induced a stronger increase in iron oxidic species size with decreasing iron loading. Independent of the precursor, thick layer calcination yielded higher aggregated and larger iron oxidic species on SBA-15 support material compared to thin layer calcination.

All samples possessed edge energy values higher than that of crystalline α-Fe_2O_3 (2.10 eV), as observed for the mechanical mixture. Hence, for all Fe_xO_y/SBA-15_Th samples, iron oxidic species on SBA-15 were smaller than crystallites in α-Fe_2O_3. Furthermore, edge energies of Fe_xO_y/SBA-15_Th samples were lower than 4.68 eV being the edge energy of isolated tetrahedrally coordinated Fe^{3+} in Fe-MFI zeolite (Si/Fe = 64). [116] As third reference, Fe(III) nitrate precursor consisting of isolated octahedrally coordinated Fe^{3+} species was used and edge energy was determined to 3.53 eV. Except for 2.5wt% Fe_Citrate_Th, all Fe_xO_y/SBA-

15_Th samples possessed edge energy values lower than 3.53 eV. Thus, supported iron oxidic species were not only larger than isolated tetrahedrally coordinated Fe^{3+} species, but also larger than isolated octahedrally coordinated Fe^{3+} species, except for 2.5wt% Fe_Citrate_Th. A decrease in edge energy with increasing iron loading for all Fe_xO_y/SBA-15_Th samples correlated with an increasing iron oxidic species size. Furthermore, higher edge energies of citrate_Th samples compared to those of nitrate_Th samples indicated smaller iron oxidic species resulting from citrate precursor. Accordingly, precursor-dependent and iron loading-dependent particle size effects were observable for supported iron oxidic species on SBA-15 independent of layer thickness during calcination.

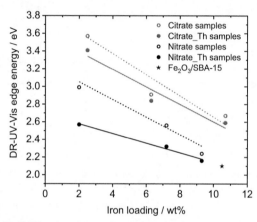

Figure 6.10: DR-UV-Vis edge energy of Fe_xO_y/SBA-15_Th, corresponding Fe_xO_y/SBA-15 samples, and the mechanical mixture Fe_2O_3/SBA-15 as function of iron loading.

In order to differentiate between overlapping absorption bands, DR-UV-Vis spectra of Fe_xO_y/SBA-15_Th catalysts were fitted with Gaussian functions. Fitting procedure was conducted analogously to that of Fe_xO_y/SBA-15 samples as described in chapter 5.3.3. The position of the fitted Gaussian functions, and therefore, position of the absorption bands varied dependent on layer thickness during calcination. This indicated differences in iron oxidic species supported on SBA-15, which was corroborated by variations in color of Fe_xO_y/SBA-15 samples dependent on layer thickness during calcination (Figure 6.1). For 9.3wt% Fe_Nitrate_Th, a fifth Gaussian function was required to describe the broad absorption in DR-UV-Vis spectrum, for all other Fe_xO_y/SBA-15_Th samples, three Gaussian functions were sufficient. Variations in positions of Gaussian functions in DR-UV-Vis spectra of Fe_xO_y/SBA-15_Th compared to corresponding Fe_xO_y/SBA-15 samples are clarified in Figure 6.11.

Only highest loaded nitrate_Th sample, 9.3wt% Fe_Nitrate_Th, showed a weak absorption band at wavenumbers lower than 20000 cm^{-1}, assigned to d-d transitions in an octahedral coordination sphere (Figure 6.9). This indicated the presence of large iron oxide aggregates with a geometry similar to that of α-Fe_2O_3. [118, 119]

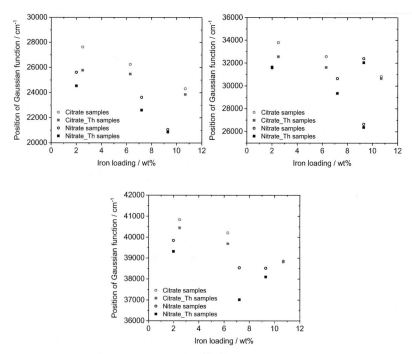

Figure 6.11: Position of Gaussian functions in experimental DR-UV-Vis spectra of Fe_xO_y/SBA-15_Th and corresponding Fe_xO_y/SBA-15 samples as function of iron loading. Only Gaussian functions attributed to CT transitions are depicted.

Thick layer calcination induced a red-shift of all Gaussian functions, indicative of an increased iron oxidic species size in Fe_xO_y/SBA-15_Th compared to Fe_xO_y/SBA-15 samples. In DR-UV-Vis spectra of Fe_xO_y/SBA-15 samples, only for 10.7wt% Fe_Citrate, 9.3wt% Fe_Nitrate, and 7.2wt% Fe_Nitrate, an absorption band at wavenumbers lower than 25000 cm^{-1} was observed. Absorption bands at wavenumbers lower than 25000 cm^{-1} were attributed to CT transitions in α-Fe_2O_3 particles. [48, 50, 112] Conversely, after thick layer calcination, all nitrate_Th samples and highest loaded citrate_Th sample showed an absorption band resulting from CT transitions in α-Fe_2O_3 particles. Hence, for citrate_Th samples with iron loadings lower than 10.7wt%, thick layer calcination did not result in small fractions of α-Fe_2O_3 particles. For all Fe_xO_y/SBA-15_Th samples, the Gaussian function in wavenumber range between 25000 and 33333 cm^{-1} was attributed to CT transitions in small Fe_xO_y oligomers. [48, 50, 112, 113] For 9.3wt% Fe_Nitrate_Th, two different Gaussian functions, ascribed to CT transitions in small Fe_xO_y oligomers, were required. This indicated a bimodal particle size distribution as observed for 9.3wt% Fe_Nitrate (see chapter 5.3.3). Gaussian function located at wavenumbers higher than 33333 cm^{-1} but lower than 41000 cm^{-1} was assigned to charge transfer transitions of isolated octahedrally coordinated Fe^{3+} species. [48, 112, 117] Accordingly, despite increased species size due to thick layer calcination, Fe_xO_y/SBA-15_Th samples still contained a small fraction of isolated octahedrally coordinated Fe^{3+} species. Partially isolated iron oxidic species in the pore system of SBA-15 were consistent with results from N_2 physisorption and TEM measurements. Precursor-dependent particle size effect, i.e. higher wavenumber ranges of Gaussian functions for Fe_xO_y/SBA-15 samples obtained from citrate precursor, remained unaffected by layer thickness during calcination. Despite the red-shift of the absorption, overall shape of DR-UV-Vis spectra of Fe_xO_y/SBA-15_Th, except for 9.3wt% Fe_Nitrate_Th, remained similar to those of corresponding Fe_xO_y/SBA-15 samples, while differing distinctly from that of the mechanical mixture (Figure 5.14). Hence, significant amounts of large crystalline α-Fe_2O_3 particles in Fe_xO_y/SBA-15_Th samples were excluded. Apparently, thick layer calcination still yielded small and dispersed iron oxidic species in the pore system of SBA-15.

6.3.4 Correlation between Fe_xO_y species size and powder layer thickness during calcination

As described in pervious chapters, powder layer thickness during calcination affected structural characteristics of Fe_xO_y/SBA-15 catalysts. Fe_xO_y species size of all Fe_xO_y/SBA-15 catalysts increased due to thick layer calcination. During thick powder layer calcination, retention time of gaseous decomposition products of the Fe(III) precursors was extended compared to thin powder layer calcination. Accordingly, a higher concentration of these gaseous decomposition products in the powder layer resulted during calcination. Due to this higher concentration of gaseous decomposition products of the precursors, gas phase during calcination became relevant. Conversely, for thin layer calcination, gas phase was negligible due to the fast removal of gaseous decomposition products of the precursors. Moreover, the effect of powder layer thickness during calcination on Fe_xO_y/SBA-15 catalysts was dependent on the precursor. The nitrate precursor induced a more pronounced Fe_xO_y species growth during thick layer calcination compared to the citrate precursor. In principle, three effects, i.e. composition of the gas phase, concentration of gaseous decomposition products, and dispersion of the Fe(III) ions, might be relevant during thick layer calcination for precursor-dependent differences in resulting iron oxidic species size. In the following, it will be evaluated which of these three effects was decisive for the precursor-dependent differences in Fe_xO_y species size of Fe_xO_y/SBA-15_Th catalysts.

As first effect, the composition of the gas atmosphere might influence the Fe_xO_y species growth. For the nitrate precursor, Fe(III) nitrate nonahydrate, thermal decomposition during calcination yielded nitrogen oxides and H_2O as gaseous decomposition products (eq.(6.1), Figure A.12.7, [54]). Conversely, thermal decomposition of the citrate precursor, (NH$_4$, Fe(III)) citrate, resulted in nitrogen oxides, H_2O, and CO_2 as gaseous decomposition products (eq.(6.2), Figure A.12.8, [54]). For both precursors, nitrogen oxides and H_2O were observed. Therefore, these same gaseous decomposition products cannot induce the precursor-dependent differences in iron oxidic species size. Consequently, CO_2 was the only difference in gas phase composition. Hence, only CO_2 could induce the smaller and less aggregated Fe_xO_y species of the citrate_Th samples. However, it appeared hardly conceivable that CO_2 significantly influenced iron oxidic species growth. Nevertheless, a possible influence of different compositions of the gas atmosphere on resulting iron oxidic species could not be excluded.

$$Fe(NO_3)_3 \cdot 9H_2O \rightarrow Fe(NO_3)_3 \cdot 3H_2O + 6\,H_2O \rightarrow 0.5\,\alpha\text{-}Fe_2O_3 + 3\,H_2O + 3\,NO_x \qquad (6.1)$$

$$[NH_4^+]_x\,[Fe^{3+}]_y\,[C_6H_5O_7^{3-}] \rightarrow 0.5\,\alpha\text{-}Fe_2O_3 + 4.5\,H_2O + NO_x + 6\,CO_2 \qquad (6.2)$$

As second effect, the concentration of gaseous decomposition products, and consequently a shift of the equilibrium during thermal decomposition of the precursors, was considered. Thermal decomposition of the precursors during calcination in thick powder layer may be described as heterogeneous equilibrium reaction. The extended retention time of the gaseous decomposition products during thick layer calcination shifted the equilibrium to the side of the educts, according to Le Chatelier's principle. This induced an enhanced iron oxidic species growth and agglomeration. Compared to the nitrate precursor, thermal decomposition of the citrate precursor yielded higher concentrations of gaseous decomposition products (eq.(6.1) and eq.(6.2)). Accordingly, thick layer calcination of citrate samples induced a more pronounced shift of the equilibrium to the side of the educts. Therefore, a more pronounced iron oxidic species growth would be expected for the citrate samples compared to the nitrate samples. However, this was not observed. Quite the opposite, the nitrate precursor induced a more pronounced Fe_xO_y species growth during thick layer calcination compared to the citrate precursor. Hence, this effect cannot induce the precursor-dependent differences in Fe_xO_y species size. However, the increased Fe_xO_y species size of all Fe_xO_y/SBA-15 catalysts due to thick layer calcination can be explained based on this effect.

As third effect, the dispersion of the Fe(III) ions was considered. A higher dispersion of Fe(III) ions hindered Fe_xO_y species growth. Due to the different strengths of chelating effect of the two precursors, dispersion of Fe(III) ions varied dependent on the precursor. The citrate precursor possessing a more pronounced chelating effect, induced higher dispersed and encapsulated Fe(III) ions before calcination. Therefore, Fe(III) ions of the citrate precursor were too far away from each other which hindered aggregation during thick layer calcination. Accordingly, Fe_xO_y species growth and aggregation of citrate_Th samples was less pronounced compared to that of nitrate_Th samples. Apparently, this effect was stronger than that of the concentration of gaseous decomposition products during thermal decomposition of the precursors. Furthermore, an effect of the dispersion of Fe(III) ions appeared more reasonable than an effect of the composition of the gas atmosphere, i.e. presence of CO_2. Hence, the dispersion of the Fe(III) ions was decisive for the precursor-dependent differences in Fe_xO_y species size of Fe_xO_y/SBA-15_Th catalysts.

Besides Fe(III) precursor, iron loading affected the influence of layer thickness during calcination on resulting iron oxidic species. Independent of the precursor, lowest loaded Fe_xO_y/SBA-15_Th catalysts showed the most increased iron oxidic species size compared to corresponding Fe_xO_y/SBA-15 catalysts. This iron loading-dependent influence on Fe_xO_y species growth and aggregation was surprising. Since all above described effects were precursor-dependent, none of them could be assigned to the most increased Fe_xO_y species size of lowest loaded Fe_xO_y/SBA-15_Th catalysts. Apparently, the extended retention time of gaseous decomposition products was more influential for low iron loadings up to 2.5wt%. A possible explanation might be a dependence of the retention time of gaseous decomposition products during calcination on the amount of precursor molecules, and therefore, on iron loading. At low amounts of precursor molecules, retention time was significantly shorter compared to that for higher amounts of precursor molecules. The difference in retention time of gaseous decomposition products between thick and thin layer calcination was presumably higher for low amounts of precursor molecules. Consequently, a more enhanced iron oxidic species growth resulted for lowest loaded Fe_xO_y/SBA-15_Th catalysts.

6.4 Reducibility of Fe_xO_y/SBA-15_Th catalysts

6.4.1 Temperature-programmed reduction with hydrogen

TPR traces of Fe_xO_y/SBA-15_Th catalysts measured during reduction with 5% H_2 in argon at a heating rate of 10 K/min are depicted in Figure 6.12 and Figure 6.13. Differences in reduction profiles of citrate_Th compared to citrate samples were only minor (Figure 6.12 and Figure 5.27). Lowest loaded citrate_Th sample possessed one single reduction peak indicating a single-step reduction mechanism. The reduction peak of this citrate_Th sample was slightly broadened compared to that of the citrate sample. Higher loaded citrate_Th samples showed a two-step reduction mechanism (Figure 6.12). For all citrate_Th samples, a slight shift of the first TPR maximum towards lower temperatures compared to corresponding citrate samples was observed. This shift of the first TPR maximum indicated a better reducibility of the citrate_Th samples. This improved reducibility of citrate_Th samples was attributed to higher amounts of Fe_xO_y oligomers as well as higher aggregated iron oxidic species induced by thick layer calcination.

In contrast to citrate_Th, nitrate_Th samples showed a more pronounced influence of layer thickness during calcination on reduction mechanism. TPR traces of nitrate_Th samples are depicted in Figure 6.13. While lowest loaded nitrate sample possessed a single-step reduction mechanism (Figure 5.28), corresponding nitrate_Th sample possessed a two-step reduction mechanism. Thick layer calcination of the higher loaded nitrate samples also induced an increasing complexity of the reduction mechanism. TPR traces of 7.2wt% Fe_Nitrate_Th and 9.3wt% Fe_Nitrate_Th showed a multi-step reduction mechanism with less resolved reduction peaks in the lower temperature range. For nitrate_Th samples, a shift of the first TPR maximum to higher temperatures was observed. This shift of the first TPR maximum for the nitrate_Th samples was probably induced by an increasing amount of crystalline α-Fe_2O_3 with increasing iron loading. For crystalline α-Fe_2O_3, increasing particle size correlates with a hindered reducibility. [49, 50, 171] In addition to the shift of the TPR maxima, a significant broadening of the first reduction peak for nitrate_Th samples was observed. This broadening was attributed to both broadened size distribution of iron oxidic species and bulk diffusion limitation due to higher amounts of crystalline α-Fe_2O_3. [49]

Additionally, reduction degree α traces were extracted by integration of TPR traces of all Fe_xO_y/SBA-15_Th samples. Resulted α traces are depicted in the inset in Figure 6.12 and Figure 6.13 for citrate_Th and nitrate_Th samples, respectively. All Fe_xO_y/SBA-15_Th samples possessed characteristic sigmoidal α traces as expected for non-isothermal reduction conditions. 6.3wt% Fe_Citrate_Th and 10.7wt% Fe_Citrate_Th showed α traces being divided into two sigmoidal parts, complying with a distinct two-step reduction mechanism as indicated by TPR profiles. α traces of all nitrate_Th samples showed characteristic sigmoidal behavior with an inflection at high temperatures. This inflection in α traces of all nitrate_Th samples correlated with a multi-step reduction mechanism with less resolved reduction peaks in TPR profiles.

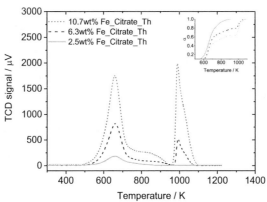

Figure 6.12: TPR traces of citrate_Th samples at various iron loadings measured in 5% H_2 in argon at 10 K/min. Inset depicts reduction degree α traces from left to right with increasing iron loading.

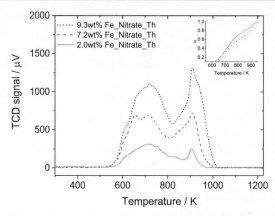

Figure 6.13: TPR traces of nitrate_Th samples at various iron loadings measured in 5% H_2 in argon at 10 K/min. Inset depicts reduction degree α traces from left to right with increasing iron loading.

6.4.2 Solid-state kinetic analysis of Fe_xO_y/SBA-15_Th under non-isothermal conditions

Introduction

As shown for Fe_xO_y/SBA-15 catalysts (see chapter 5.4.3), solid-state kinetic analysis can be successfully applied to supported systems, and therefore, constitutes an additional analysis method for characterizing supported species. Accordingly, solid-state kinetic analysis methods were applied to Fe_xO_y/SBA-15_Th samples. Hereby, the influence of layer thickness during calcination on reducibility of supported iron oxidic species on SBA-15 dependent on both precursor and iron loading was elucidated. Therefore, TPR measurements in 5% H_2 in argon were conducted at varying heating rate. Prior to applying model-independent and model-dependent solid-state kinetic analysis methods, TPR traces of Fe_xO_y/SBA-15_Th were transformed to reduction degree α traces. TPR traces and reduction degree α traces, as function of temperature at various heating rates are depicted for sample 6.3wt% Fe_Citrate_Th (Figure 6.14). Both TPR traces and reduction degree α traces were shifted to higher temperatures with increasing heating rate, β. This was observed for all Fe_xO_y/SBA-15_Th samples. In the following, a detailed solid-state kinetic analysis of the reduction traces will be presented. Fundamentals of applied solid-state kinetic analysis methods are described in chapter 5.4.3.

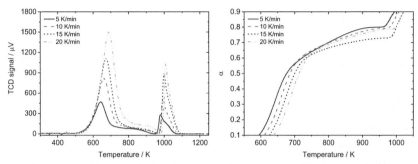

Figure 6.14: TPR traces (left) and reduction degree α traces (right) as function of temperature for reduction (5% H_2 in argon) of 6.3wt% Fe_Citrate_Th at various heating rates (5, 10, 15, and 20 K/min).

Kissinger method

Model-independent Kissinger method yielded a single apparent activation energy, E_a, of rate-determining step during reduction of Fe_xO_y/SBA-15_Th samples. By depicting $ln[\beta/T_m^2]$ as function of $1000/T_m$, apparent activation energy of the reduction of Fe_xO_y/SBA-15_Th was calculated from the slope of the resulting straight line. [82, 87] In order to enable comparison of E_a for all samples despite differing TPR traces, T_m corresponded to the first maximum of the TPR traces (Figure 6.12 and Figure 6.13). Apparent activation energies for Fe_xO_y/SBA-15_Th samples, corresponding Fe_xO_y/SBA-15 samples, and the mechanical mixture are summarized in Table 6.4. As already observed for Fe_xO_y/SBA-15 catalysts, used precursor significantly influenced reducibility, and hence, apparent activation energy of rate-determining step in reduction of Fe_xO_y/SBA-15_Th catalysts. Highest apparent activation energy of rate-determining step in reduction resulted for lowest loaded nitrate_Th sample. Increasing iron loading within the nitrate_Th samples correlated with a decreasing apparent activation energy of reduction. Higher loaded nitrate_Th samples, 7.2wt% Fe_Nitrate_Th and 9.3wt% Fe_Nitrate_Th, exhibited the same value of E_a within experimental errors as the mechanical mixture of α-Fe_2O_3 and SBA-15. This might be indicative of a similar size of iron oxidic species. Similarities in iron oxidic species size were corroborated by results from DR-UV-Vis spectroscopy and XRD. 7.2wt% Fe_Nitrate_Th and 9.3wt% Fe_Nitrate_Th possessed a DR-UV-Vis edge energy of 2.3 respectively 2.2 eV, being similar to that of Fe_2O_3/SBA-15 (2.1 eV) (Figure 6.10). Furthermore, these samples showed XRD peaks corresponding to crystalline α-Fe_2O_3 (Figure 6.7). Accordingly, increasing amount of crystalline α-Fe_2O_3 agreed with the decrease in E_a for higher loaded nitrate_Th samples. The highest apparent activation energy of 2.0wt% Fe_Nitrate_Th coincided with the most distinct broadening of the first TPR peak (Figure 6.13). Thick layer calcination of the lowest loaded nitrate sample affected the most pronounced difference in iron oxidic species size. For this sample, increased species size, broadened species size distribution, and hence, increased complexity of the reduction mechanism were observed. Conversely, using the citrate precursor in thick layer calcination yielded highest apparent activation energy in reduction for highest loaded samples, while lowest loaded citrate_Th sample possessed lowest E_a. Furthermore, values of E_a for citrate_Th samples were higher compared to those of nitrate_Th samples for iron loadings higher than 2.5wt%. This was indicative of a precursor-dependent trend in apparent activation energy as function of iron loading for Fe_xO_y/SBA-15_Th catalysts, as already observed for Fe_xO_y/SBA-15 catalysts (see chapter 5.4.3).

For lowest loaded Fe_xO_y/SBA-15_Th samples, the most pronounced influence of layer thickness during calcination on apparent activation energy in reduction was observed (Figure 6.15). A significant increase in apparent activation energy of the reduction of 2.5wt% Fe_Citrate_Th and 2.0wt% Fe_Nitrate_Th, compared to corresponding Fe_xO_y/SBA-15 samples, was observed. This might be induced by a broadened species size distribution, being reflected in the broadening of the first TPR maxima. Accordingly, reduction of iron oxidic species with different sizes, ascribed to the first TPR peak, was more difficult compared to that of iron oxidic species with a narrow species size distribution, yielding an increased apparent activation energy.

Table 6.4: Apparent activation energy, E_a, of rate-determining step in reduction of Fe_xO_y/SBA-15_Th, corresponding Fe_xO_y/SBA-15 samples, and mechanical mixture Fe_2O_3/SBA-15. E_a values were determined by Kissinger method and reduction was conducted in 5% H_2 in argon.

	E_a / kJ/mol	
	Thick layer calcined	Thin layer calcined
2.5wt% Fe_Citrate	76 ± 7	39 ± 8
6.3wt% Fe_Citrate	102 ± 8	113 ± 8
10.7wt% Fe_Citrate	88 ± 8	104 ± 8
2.0wt% Fe_Nitrate	130 ± 3	88 ± 8
7.2wt% Fe_Nitrate	69 ± 5	84 ± 1
9.3wt% Fe_Nitrate	70 ± 5	62 ± 8
Fe_2O_3/SBA-15	59 ± 7	

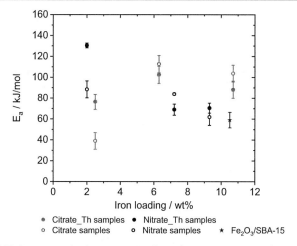

Figure 6.15: Apparent activation energy, E_a, of rate-determining step in reduction (5% H_2 in argon) determined by Kissinger method for Fe_xO_y/SBA-15_Th, corresponding Fe_xO_y/SBA-15 samples, and the mechanical mixture as function of iron loading.

Method of Ozawa, Flynn, and Wall

Isoconversional, model-independent method of Ozawa, Flynn, and Wall (OFW) was applied for determining the evolution of apparent activation energy of rate-determining step as function of reduction degree α. [82, 88, 89, 150] Fundamentals of OFW method are described in chapter 5.4.3. The evolution of apparent activation energy, $E_{a,\alpha}$, as function of reduction degree α was determined for all Fe_xO_y/SBA-15_Th samples. Therefore, in a first step, temperatures $T_{\alpha,\beta}$ for defined reduction degrees α were determined from the experimental α traces at various heating rates. Based on TPR traces, only temperatures corresponding to the first reduction peak were considered. Consequently, temperatures $T_{\alpha,\beta}$ were determined for limited α range of 0.1-0.5 for citrate_Th samples, and for α = 0.1-0.4 for nitrate_Th samples. For all Fe_xO_y/SBA-15_Th catalysts, $\Delta\alpha$ was 0.05. Afterwards, decade logarithm of the heating rate as function of $1000/T_{\alpha,\beta}$ for the different reduction degrees was calculated according to equation (5.23). Linear regression of the resulting straight lines resulted in $E_{a,\alpha}$, which was corrected according to Senum-Yang. [82, 150, 153] Apparent activation energy as function of reduction degree α together with apparent activation energy determined by Kissinger method for citrate_Th samples is depicted in Figure 6.16.

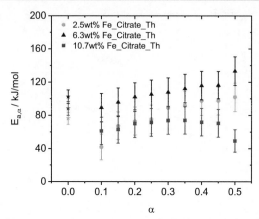

Figure 6.16: Apparent activation energy, $E_{a,\alpha}$, as function of reduction degree, α, for reduction of citrate_Th samples in 5% H_2 in argon according to Senum-Yang approximation. Apparent activation energies determined by Kissinger method are depicted at $\alpha = 0$.

For all citrate_Th samples, apparent activation energy obtained from OFW method agreed with that obtained from Kissinger method. Apparent activation energy $E_{a,\alpha}$ for reduction of citrate_Th samples showed no significant variations in the α range within the error limits. This was indicative of a single-step reduction mechanism for citrate_Th samples in α range up to 0.5, coinciding with the sharp single reduction peak in the TPR traces of these samples (Figure 6.12). Hence, reduction mechanism of FexOy/SBA-15 samples obtained from citrate precursor was unaffected by layer thickness during calcination.

The evolution of apparent activation energy as function of reduction degree α for nitrate_Th samples differed from that of citrate_Th samples. (Figure 6.17). Samples 9.3wt% Fe_Nitrate_Th and 2.0wt% Fe_Nitrate_Th showed an increasing apparent activation energy $E_{a,\alpha}$ with increasing reduction degree α. The increase of $E_{a,\alpha}$ was less pronounced for 7.2wt% Fe_Nitrate_Th. This may indicate a change in rate-determining step during a multi-step reduction mechanism for all nitrate_Th samples. [155, 156] A multi-step reduction mechanism coincided with the observed broad and less resolved first reduction peak in low temperature range in TPR traces of all nitrate_Th samples (Figure 6.13).

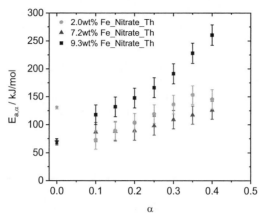

Figure 6.17: Apparent activation energy, $E_{a,\alpha}$, as function of reduction degree, α, for reduction of nitrate_Th samples in 5% H_2 in argon according to Senum-Yang approximation. Apparent activation energies determined by Kissinger method are depicted at $\alpha = 0$.

Coats-Redfern method

In addition to model-independent Kissinger and OFW methods, model-dependent Coats-Redfern method [157] was applied for providing a complementary analysis of non-isothermal kinetic data of Fe_xO_y/SBA-15_Th catalysts. Fundamentals of Coats-Redfern method are described in chapter 5.4.3. Plotting $ln[g(\alpha)/T^2]$ as function of reciprocal temperature results in straight lines for suitable solid-state reaction models. Apparent activation energies can be obtained from the slope of resulting straight lines. [149, 157] Only solid-state reaction models, $g(\alpha)$, resulting in similar apparent activation energies as obtained from model-independent methods as well as good linear regressions were selected for further analysis. Besides nucleation models, including power law models (P) and Avrami-Erofeyev models (A), the autocatalytic Prout-Tompkins model (B1), diffusion models (D), geometrical contraction models (R), and reaction order-based models (F) were tested. Analysis procedure according to Coats-Redfern method for Fe_xO_y/SBA-15_Th catalysts was performed analogously to that described for Fe_xO_y/SBA-15 catalysts (see chapter 5.4.3). For all citrate_Th samples, F1, D2, A3, and R2 solid-state reaction models revealed wide linear ranges by plotting $ln[g(\alpha)/T^2]$ as function of reciprocal temperature. Linear ranges of reduction degree α curves using the F1 solid-state reaction model for 2.5wt% Fe_Citrate_Th are depicted in Figure 6.18, left. As observed for 2.5wt% Fe_Citrate (Table 5.11), apparent activation energies at different heating rates for diffusion and Avrami-Erofeyev reaction models, i.e. D2 and A3 reaction models, were significantly higher or lower compared to the results of Kissinger and OFW methods. Hence, neither the D2 nor the A3 solid-state reaction model were considered for further analysis. The R2 reaction model is described as geometrical contraction model in which nucleation occurs on the surface of the cylindrical crystal. Thus, the reaction rate is determined by the decreasing interface area between reactant and product phase during reaction. [149] This concept appeared hardly applicable to citrate_Th samples with small and dispersed iron oxidic species located in the pore system of SBA-15. For the F1 solid-state reaction model, not only values of apparent activation energy agreed with those obtained from Kissinger and OFW methods, but also underlaying concept was well applicable to small and dispersed iron oxidic species of citrate_Th samples. Consequently, the F1, Mampel, model was chosen as suitable solid-state reaction model for 2.5wt% Fe_Citrate_Th and all other citrate_Th samples. For citrate samples, thick layer calcination did not lead to a change in suitable solid-state reaction model.

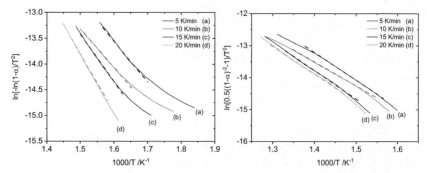

Figure 6.18: Linear ranges of reduction degree α curves using the F1 solid-state reaction model for sample 2.5wt% Fe_Citrate_Th (left, straight lines) and those using the F3 solid-state reaction model for sample 7.2wt% Fe_Nitrate_Th (right, straight lines). Reduction was conducted in 5% H_2 in argon. Dashed lines: Linear regressions.

For lowest loaded nitrate_Th sample, analysis procedure according to Coats-Redfern method also resulted in F1 solid-state reaction model being suitable for describing the reduction. Hence, thick layer calcination of 2.0wt% Fe_Nitrate induced no change in suitable solid-state reaction model. Conversely, for higher loaded nitrate_Th samples, F1, D1, R3, and F3 solid-state reaction models revealed wide linear ranges by plotting $ln[g(\alpha)/T^2]$ as function of reciprocal temperature. Linear ranges of reduction degree α curves using the F3 solid-state reaction model for 7.2wt% Fe_Nitrate_Th are depicted in Figure 6.18, right. Apparent activation energies for solid-state reaction models F1, D1, R3, and F3 obtained from the slope of resulting straight lines are given in Table 6.5.

Table 6.5: Apparent activation energy, E_a, of reduction of 7.2wt% Fe_Nitrate_Th in 5% H_2 in argon at various heating rates depending on the applied solid-state reaction model.

Heating rate /	E_a / kJ mol^{-1}			
K min^{-1}	D1	R3	F1	F3
5	122.4 ± 0.3	59.1 ± 0.2	58.8 ± 0.1	74.2 ± 0.1
10	135.9 ± 0.6	66.7 ± 0.3	58.6 ± 0.4	68.5 ± 0.2
15	141.0 ± 0.8	69.0 ± 0.4	64.7 ± 0.3	77.2 ± 0.2
20	144.2 ± 0.9	74.6 ± 0.3	59.1 ± 0.3	75.8 ± 0.2

For the D1 solid-state reaction model, apparent activation energies of reduction at different heating rates were significantly higher compared to the results from Kissinger and OFW methods. Consequently, the D1 was not further considered as suitable reaction model for the reduction of 7.2wt% Fe_Nitrate_Th. Moreover, the F1 solid-state reaction model yielded significantly lower values of apparent activation energy compared to those resulted from model-independent solid-state kinetic analysis methods. This indicated that thick layer calcination of 7.2wt% Fe_Nitrate induced a change in suitable solid-state reaction model. Therefore, first-order reaction model, F1, was no longer suitable for describing the reduction of 7.2wt% Fe_Nitrate_Th. Apparent activation energies which resulted from the R3 reaction model were also lower than those from Kissinger and OFW methods. Besides lower values of apparent activation energy, R3 reaction model appeared hardly applicable to the reduction of 7.2wt% Fe_Nitrate_Th regarding the underlying assumptions of this reaction model. The R3 reaction model is denoted as contracting sphere model with nucleation occurring rapidly on the surface of the particles. Geometrical contraction models (R) assume a reaction interface progressing towards the center of the crystal particle during reaction. However, Fe(III) species in 7.2wt% Fe_Nitrate_Th were small and located in the pore system of SBA-15. Only a minor fraction of higher aggregated iron oxidic species and crystalline α-Fe$_2$O$_3$ was observed for this sample, excluding the R3 as suitable reaction model. Accordingly, third-order, F3, solid-state reaction model was chosen as suitable reaction model for the reduction of 7.2wt% Fe_Nitrate_Th. Analysis procedure for 9.3wt% Fe_Nitrate_Th also resulted in F3 reaction model being suitable for describing the reduction mechanism of this sample. For third-order reaction model, equation (5.25) results in

$$\frac{d\alpha}{dt} = k(1 - \alpha)^3 \tag{6.3}$$

with rate of reaction, $d\alpha/dt$, rate constant, k, and reduction degree, α. [149, 158]
The integral expression $g(\alpha)$ for third-order reaction model can be obtained after separating variables and integrating and is expressed by

$$g(\alpha) = \frac{1}{2}[(1 - \alpha)^{-2} - 1]. \tag{6.4}$$

The increased reaction order for higher loaded nitrate_Th samples was correlated with an increased amount of higher aggregated iron oxidic species and the presence of a minor fraction of crystalline α-Fe$_2$O$_3$.

JMAK kinetic analysis

Johnson-Mehl-Avrami-Kolmogorov (JMAK) kinetic analysis was applied to geometrically describe the reduction of Fe_xO_y/SBA-15_Th samples under non-isothermal conditions. [160–162] Fundamentals of JMAK kinetic analysis are explained in chapter 5.4.3. JMAK kinetic analysis procedure for Fe_xO_y/SBA-15_Th samples was performed analogously to that for Fe_xO_y/SBA-15 samples (see chapter 5.4.3). Topological dimension and Avrami exponent as function of temperature and heating rate for all Fe_xO_y/SBA-15_Th samples are depicted in Figure 6.19. Since thick layer calcination induced larger iron oxidic species, JMAK kinetic analysis became applicable to 2.5wt% Fe_Citrate_Th. This was not possible for 2.5wt% Fe_Citrate (see chapter 5.4.3). For all citrate_Th samples, topological dimension and Avrami exponent were approximately one (Figure 6.19, left), as observed for the citrate samples. Accordingly, for Fe_xO_y/SBA-15 samples obtained from citrate precursor, thick layer calcination influenced iron oxidic species sizes, while geometrical characteristics of the reduction were unaffected. Furthermore, for all citrate_Th samples, first-order, Mampel, solid-state reaction model was identified as being suitable for the reduction. Mampel solid-state reaction model was consistent with the reduction being governed by site saturation ($n = m = 1$). Hence, results from JMAK kinetic analysis and model-dependent Coats-Redfern method agreed well for all citrate_Th samples.

JMAK kinetic analysis of nitrate_Th samples revealed an influence of layer thickness during calcination on geometrical characteristics of the reduction. For sample 2.0wt% Fe_Nitrate_Th, Avrami exponent and topological dimension were approximately one, indicating reduction being governed by site saturation, similar to the citrate_Th samples. This coincided with the first-order, Mampel, solid-state reaction model being suitable to describe the reduction of 2.0wt% Fe_Nitrate_Th. An Avrami exponent of one and a topological dimension of 1.5 was obtained for sample 7.2wt% Fe_Nitrate_Th. This increase in topological dimension compared to the lowest loaded nitrate_Th sample and all citrate_Th samples was correlated with both increased iron oxidic species size and increased amount of crystalline α-Fe_2O_3 (see chapter 6.3.3). Similar to 7.2wt% Fe_Nitrate_Th, 9.3wt% Fe_Nitrate_Th possessed an Avrami exponent of one and a topological dimension of 1.5. Topological dimension of this highest loaded nitrate_Th sample was slightly decreased after thick layer calcination compared to corresponding nitrate sample. This decrease in topological dimension, from 2 to 1.5, was ascribed to the increased degree of micropore filling and the smoother SBA-15 surface due to larger iron oxidic species resulting from thick layer calcination. Despite an increased amount of both higher aggregated iron oxidic species and crystalline α-Fe_2O_3 for 7.2wt% Fe_Nitrate_Th and 9.3wt% Fe_Nitrate_Th, resulting iron oxidic species of these samples were smaller than crystalline α-Fe_2O_3 particles of the mechanical mixture. These differences in iron oxidic species sizes agreed with the differences in topological dimension of reduction for higher loaded nitrate_Th samples compared to the mechanical mixture ($m = 2$-3). Furthermore, increased topological dimension of reduction for 7.2wt% Fe_Nitrate_Th and 9.3wt% Fe_Nitrate_Th compared to 2.0wt% Fe_Nitrate_Th and all citrate_Th samples was consistent with the increased reaction order of solid-state reaction model, i.e. F3 solid-state reaction model being suitable for describing the reduction instead of F1.

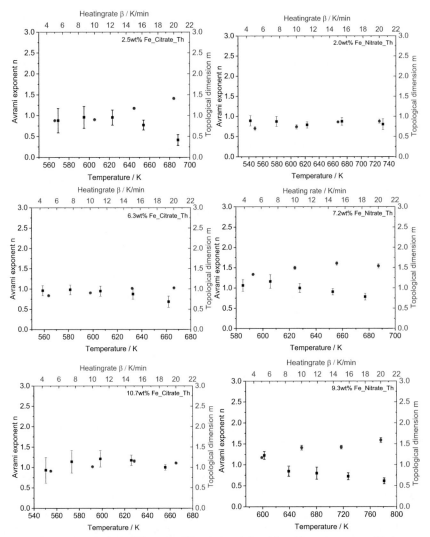

Figure 6.19: Topological dimension (blue circles, right axis) and Avrami exponent (black squares, left axis) from JMAK kinetic analysis as function of temperature and heating rate for citrate_Th (left) and nitrate_Th (right) samples. Increasing iron loading from top to bottom.

6.4.3 Correlation between solid-state kinetic analysis and structural characterization

Similar to Fe_xO_y/SBA-15 samples (see chapter 5.4.4), results from solid-state kinetic analysis of Fe_xO_y/SBA-15_Th samples agreed well with those from structural characterization. Thick layer calcination of Fe_xO_y/SBA-15 samples influenced both structural and solid-state kinetic properties of the iron oxidic species supported on SBA-15. Thick layer calcination induced an increased iron oxidic species size due to an extended retention time of gaseous decomposition products of the precursors during synthesis of Fe_xO_y/SBA-15_Th. Hence, gas phase during calcination became relevant. Compared to Fe(III) ions of the citrate precursor, those of the nitrate precursor were less dispersed before calcination. This yielded a more pronounced Fe_xO_y species growth and aggregation for nitrate_Th compared to citrate_Th samples (see chapter 6.3.4). This increased iron oxidic species size of nitrate_Th samples correlated with lower values of apparent activation energy of reduction for iron loadings higher than 2.5wt% compared to citrate_Th samples. Furthermore, within the nitrate_Th samples, increasing iron oxidic species size and increasing amount of higher aggregated iron oxidic species with increasing iron loading induced a decreasing apparent activation energy of reduction. Besides, increased amount of crystalline α-Fe_2O_3 for 7.2wt% Fe_Nitrate_Th and 9.3wt% Fe_Nitrate_Th coincided with values of apparent activation energy in reduction similar to that of the mechanical mixture, Fe_2O_3/SBA-15. JMAK kinetic analysis revealed a topological dimension of 1.5 for 7.2wt% Fe_Nitrate_Th and 9.3wt% Fe_Nitrate_Th. This increase in topological dimension in reduction of higher loaded nitrate_Th samples compared to all other Fe_xO_y/SBA-15_Th samples correlated with higher aggregated iron oxidic species. The slightly decreased topological dimension for 9.3wt% Fe_Nitrate_Th compared to that for 9.3wt% Fe_Nitrate was ascribed to an increased degree of micropore filling accompanied by a smoother SBA-15 surface due to larger iron oxidic species. Moreover, results from Coats-Redfern method agreed with those from structural characterization. For higher loaded nitrate_Th samples, significantly increased iron oxidic species size correlated with an increased reaction-order of suitable solid-state reaction model.

Compared to nitrate_Th samples, for citrate_Th samples, the influence of layer thickness during calcination on structural and solid-state kinetic properties was less pronounced. Despite increased iron oxidic species size due to thick layer calcination, resulting iron oxidic species remained significantly smaller and higher dispersed compared to those obtained from nitrate precursor (see chapter 6.3.3). Moreover, citrate_Th samples with iron loadings higher than 2.5wt% showed a more pronounced micropore filling, and therefore, a smoother SBA-15 surface compared to corresponding nitrate_Th samples. This was indicative of stronger interactions between iron oxidic species and SBA-15, which were induced by the more pronounced chelating effect of the citrate precursor during synthesis of Fe_xO_y/SBA-15_Th samples. Moreover, these stronger interactions between iron oxidic species and SBA-15 coincided with the higher values of apparent activation energy in reduction of citrate_Th samples (see chapter 5.6). Smaller iron oxidic species of citrate_Th compared to nitrate_Th samples yielded first-order, F1, solid-state reaction model being suitable to describe the reduction. Reduction of citrate_Th samples was furthermore governed by site saturation, being consistent with F1, Mampel, model as suitable solid-state reaction model. Lowest loaded nitrate_Th sample, 2.0wt% Fe_Nitrate_Th, consisting of similar iron oxidic species as citrate_Th samples, showed analogous results in solid-state kinetic analysis as citrate_Th samples. Precursor-dependent differences in reducibility and solid-state kinetic properties of Fe_xO_y/SBA-15_Th samples were predominantly correlated with precursor-dependent differences in resulting iron oxidic species size.

Thick layer calcination of lowest loaded samples induced the most distinct increase in iron oxidic species size and broadened species size distribution. This was ascribed to the highest difference in retention time of gaseous decomposition products between thick and thin layer calcination for lowest iron loadings. These differences in structural characteristics were reflected in the significantly increased apparent activation energy in reduction determined by Kissinger method for 2.5wt% Fe_Citrate_Th and 2.0wt% Fe_Nitrate_Th.

6.5   Catalytic performance in selective oxidation of propene

Catalytic performance of Fe_xO_y/SBA-15_Th catalysts in selective oxidation of propene was investigated at 653 K and at comparable propene conversions between 5 and 10%. To ensure the absence of gas phase diffusion effects in selective oxidation of propene, the Weisz-Prater criterion was applied for Fe_xO_y/SBA-15_Th catalysts, as described in chapter 5.5.2. C_{WP} remained 0.02 and was therefore indicative of neglectable gas phase diffusion effects. [164] Additionally, heat and mass transport limitations in catalytic reaction of Fe_xO_y/SBA-15_Th were excluded based on Koros-Nowak test. [165] Distribution of main products during selective oxidation of propene for Fe_xO_y/SBA-15_Th catalysts are summarized in Table 6.6. Acrolein, carbon oxides, CO and CO_2, and acetaldehyde were the main reaction products. Acrolein selectivity as function of time on stream is depicted in Figure 6.20. Significant differences in acrolein selectivity, and furthermore, in catalytic performance were observed for Fe_xO_y/SBA-15_Th catalysts dependent on the precursor. For citrate_Th samples, decreasing iron loading correlated with an increasing acrolein selectivity. Besides acrolein selectivity, selectivity towards acetaldehyde also increased at the expense of CO and CO_2 with decreasing iron loading. Accordingly, for citrate_Th samples, decreasing iron loading, and hence, decreasing iron oxidic species size induced a more selective reaction.

Nitrate_Th samples differed in catalytic performance compared to citrate_Th samples. Nitrate_Th samples showed a volcano-type behavior in selectivity towards acrolein and acetaldehyde as function of iron loading. Highest selectivity towards acrolein, accompanied by highest selectivity towards acetaldehyde and lowest selectivity towards carbon oxides, was observed for 7.2wt% Fe_Nitrate_Th. Increasing the iron loading up to 7.2wt% within the nitrate_Th samples correlated with an increasing iron oxidic species size and an increasing acrolein selectivity. Further increasing iron loading induced a significant fraction of crystalline α-Fe_2O_3 (see chapter 6.3.2), and hence, a decreased acrolein selectivity.

Figure 6.20 clarifies precursor-dependent differences in acrolein selectivity as function of time on stream. Highest acrolein selectivity resulted for 7.2wt% Fe_Nitrate_Th and lowest for 10.7wt% Fe_Citrate_Th. Samples 2.5wt% Fe_Citrate_Th, 2.0wt% Fe_Nitrate_Th, and 9.3wt% Fe_Nitrate_Th showed a comparable acrolein selectivity during 12 h on stream. Comparable acrolein selectivity for these three samples was correlated with a similar contribution of micropores to entire surface area, and further, a similar SBA-15 surface roughness (Table 6.2).

Table 6.6: Distribution of main products during selective oxidation of propene for Fe_xO_y/SBA-15_Th catalysts. Catalytic measurements were conducted in 5% propene, 5% oxygen in helium at 653 K. Acro: acrolein, Acet: acetaldehyde.

	Iron loading / wt%	Selectivity / %			
		Acro	Acet	CO	CO$_2$
	2.5	28.6	9.2	20.6	39.2
Citrate_Th	6.3	13.5	3.8	26.9	54.8
	10.7	11.7	2.7	28.3	56.5
	2.0	27.1	4.7	17.4	49.5
Nitrate_Th	7.2	32.0	5.0	17.3	45.3
	9.3	26.6	4.3	19.9	48.8

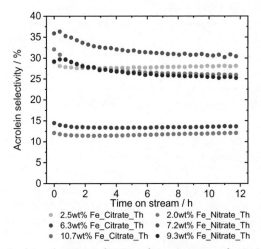

Figure 6.20: Acrolein selectivity as function of time on stream for Fe_xO_y/SBA-15_Th catalysts during selective oxidation of propene (5% propene, 5% oxygen in helium, 653 K).

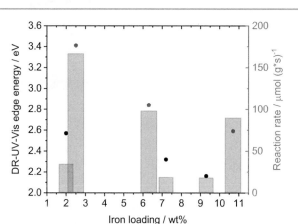

Figure 6.21: DR-UV-Vis edge energy (left axis) of citrate_Th (orange dots) and nitrate_Th (blue dots) samples and reaction rate of propene conversion (right axis, bars) as function of iron loading.

To further investigate the influence of both precursor and iron loading on catalytic performance of Fe_xO_y/SBA-15_Th samples, DR-UV-Vis edge energy and reaction rate of propene conversion as function of iron loading are compared in Figure 6.21. Independent of the precursor, reaction rate of propene conversion decreased with increasing iron loading. Consequently, increasing iron oxidic species size, expressed by decreasing DR-UV-Vis edge energy, correlated with a decreasing reaction rate of propene conversion, ascribed to a decreasing catalytic activity. Significantly smaller iron oxidic species obtained from citrate precursor compared to those obtained from nitrate precursor, coincided with significantly higher reaction rates of propene conversion for citrate_Th samples. This was indicative of a particle size effect accounting for decreasing reaction rate of propene conversion, and hence, for decreasing catalytic activity of Fe_xO_y/SBA-15_Th samples with increasing iron loading.

For citrate_Th samples, increasing acrolein selectivity with decreasing iron loading correlated with decreasing iron oxidic species size. A more detailed correlation between acrolein selectivity and surface and porosity characteristics of citrate_Th samples was elucidated based on Figure 6.22. Increasing iron loading involved a decreasing contribution of micropores to entire surface area, and hence, a higher degree of micropore filling. This was accompanied by higher ΔD_f values indicating a smoother SBA-15 surface. The degree of micropore filling was correlated with the strength of interactions between iron oxidic species and support material (see chapter 5.6). Accordingly, for Fe_xO_y/SBA-15_Th samples obtained from citrate precursor, the strength of interactions between iron oxidic species and SBA-15 determined the catalytic performance.

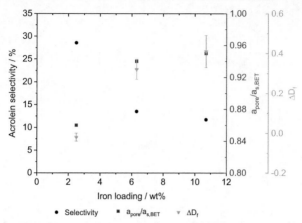

Figure 6.22: Correlation between acrolein selectivity (left axis, black dots), contribution of micropores to entire surface area, $a_{pore}/a_{s,BET}$, (right axis, blue squares), and surface roughness of SBA-15, ΔD_f, (outer right axis, green triangles) for citrate_Th samples as function of iron loading.

Differences in catalytic performance of citrate samples dependent on layer thickness during calcination were also dependent on iron loading (Figure 6.23). Differences in acrolein selectivity correlated with differences in strengths of interactions between iron oxidic species and SBA-15, expressed by ratio of $a_{pore}/a_{s,BET}$ (Table 6.2, Figure 6.24). For lowest iron loading of 2.5wt%, acrolein selectivity of citrate_Th sample was significantly increased compared to the citrate sample. This was ascribed to weaker interactions between iron oxidic species and SBA-15 compared to 2.5wt% Fe_Citrate. Conversely, for higher loaded citrate_Th samples, stronger interactions between iron oxidic species and SBA-15 resulted in a decreased acrolein selectivity compared to corresponding citrate samples.

For lowest loaded sample, 2.5wt% Fe_Citrate, besides surface properties, iron oxide dispersion influenced catalytic performance. 2.5wt% Fe_Citrate consisted of a significant fraction of isolated octahedrally coordinated Fe^{3+} species (see chapter 5.3.3). Conversely, 2.5wt% Fe_Citrate_Th consisted of higher aggregated iron oxidic species and less isolated Fe^{3+} species, yielding an increased acrolein selectivity. This complied with a higher catalytic activity in selective oxidation reactions ascribed to dimeric or small oligomeric metal oxides compared to isolated species. [11, 50, 108, 172, 173]

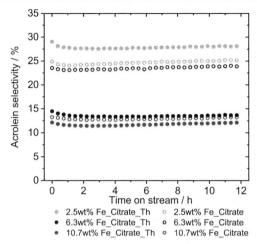

Figure 6.23: Acrolein selectivity as function of time on stream during selective oxidation of propene (5% propene, 5% oxygen in helium, at 653 K) for citrate_Th and citrate samples.

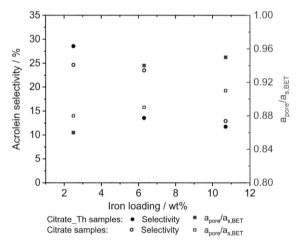

Figure 6.24: Acrolein selectivity (left axis) as function of iron loading compared to contribution of micropores to entire surface area, $a_{pore}/a_{s,BET}$, (right axis) for citrate_Th and citrate samples.

Nitrate_Th samples consisted of significantly larger, higher aggregated, and less dispersed iron oxidic species on SBA-15 compared to citrate_Th samples. Nitrate_Th samples showed an increasing acrolein selectivity with increasing iron oxidic species size up to 7.2wt% iron. As observed for nitrate samples (see chapter 5.5.6), for nitrate_Th samples, an increasing acrolein selectivity correlated with a lower apparent activation energy of rate-determining step in reduction. Accordingly, decreased apparent activation energy in reduction from 2.0wt% Fe_Nitrate_Th to 7.2wt% Fe_Nitrate_Th resulted in an increased acrolein selectivity (Figure 6.25). Despite comparable apparent activation energy in reduction for 7.2wt% Fe_Nitrate_Th and 9.3wt% Fe_Nitrate_Th, acrolein selectivity was lower for 9.3wt% Fe_Nitrate_Th. This was associated with a higher amount of less reactive crystalline α-Fe_2O_3 hindering the selective oxidation reaction.

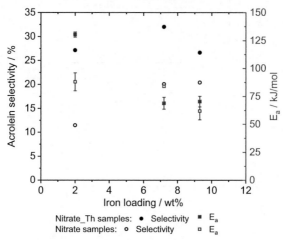

Nitrate_Th samples: • Selectivity ■ E_a
Nitrate samples: ○ Selectivity □ E_a

Figure 6.25: Acrolein selectivity (left axis) as function of iron loading compared to apparent activation energy in reduction, E_a, determined by Kissinger method (right axis) for nitrate_Th and nitrate samples.

Influence of layer thickness during calcination on catalytic performance of Fe_xO_y/SBA-15 catalysts obtained from nitrate precursor differed from that of Fe_xO_y/SBA-15 catalysts obtained from citrate precursor. Evolution of acrolein selectivity as function of time on stream during selective oxidation of propene for nitrate_Th and corresponding nitrate samples is depicted in Figure 6.26. For all iron loadings, nitrate_Th samples possessed a significantly higher acrolein selectivity compared to corresponding nitrate samples. Higher acrolein selectivity of 7.2wt% Fe_Nitrate_Th compared to 7.2wt% Fe_Nitrate agreed with the correlation between increasing iron oxidic species size, decreasing apparent activation energy in reduction, and increasing acrolein selectivity. The increased acrolein selectivity of 9.3wt% Fe_Nitrate_Th compared to 9.3wt% Fe_Nitrate was attributed to stronger interactions between iron oxidic species and SBA-15, presumably resulting in intermediate Fe-O bond strengths. For 2.0wt% Fe_Nitrate_Th, increased acrolein selectivity compared to 2.0wt% Fe_Nitrate correlated with an increased iron oxidic species size, and therefore, decreased amount of isolated octahedrally coordinated Fe^{3+} species.

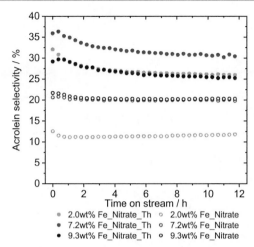

Figure 6.26: Acrolein selectivity as function of time on stream during selective oxidation of propene (5% propene, 5% oxygen in helium, at 653 K) for nitrate_Th and nitrate samples.

6.6 Interpretation of structural and solid-state kinetic properties of Fe$_x$O$_y$/SBA-15_Th catalysts and their impact on catalytic performance

Figure 6.27 depicts a schematic representation of the main structural motifs of Fe$_x$O$_y$/SBA-15_Th catalysts dependent on iron loading and precursor. Based on this schematic representation, differences in reducibility and catalytic performance of Fe$_x$O$_y$/SBA-15_Th catalysts might be interpreted. Estimation of the amount of β-FeOOH-like and α-Fe$_2$O$_3$-like species in Fe$_x$O$_y$/SBA-15_Th catalysts was not possible due to the lack of X-ray absorption measurements. Nevertheless, formation of Fe-OH structure units due to the presence of β-FeOOH-like species might be assumed (Figure 6.27, (A), (B)). Presence of β-FeOOH-like and α-Fe$_2$O$_3$-like species in Fe$_x$O$_y$/SBA-15_Th catalysts seemed likely, since differences in phase composition of Fe$_x$O$_y$/SBA-15 catalysts were ascribed to the different strengths of chelating effect of the two precursors. Furthermore, assumed presence of β-FeOOH-like and α-Fe$_2$O$_3$-like species in Fe$_x$O$_y$/SBA-15_Th catalysts was corroborated by the differences in color of the samples (Figure 6.1). As observed for Fe$_x$O$_y$/SBA-15 catalysts, it is assumed that the amount of β-FeOOH-like species decreased with increasing iron loading. Furthermore, a higher amount of β-FeOOH-like species was presumably obtained from the citrate precursor.

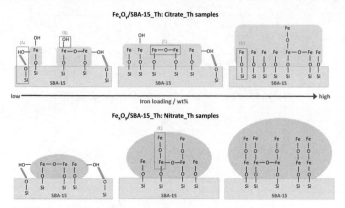

Figure 6.27: Schematic representation of structural differences of iron oxidic species supported on SBA-15 dependent on iron loading and precursor. Only the main structural motifs of Fe_xO_y/SBA-15_Th catalysts are depicted.

Increasing iron loading within citrate_Th samples correlated with an increasing degree of micropore filling. This might be induced by an increased amount of Fe-O-Si bonds (Figure 6.27, (D)), and furthermore, by an increased size of Fe_xO_y oligomers. Moreover, amount of Fe-O-Fe structure units (Figure 6.27, (C)) increased, while that of Fe-OH structure units (Figure 6.27, (A), (B)) decreased with increasing iron loading. Lower reducibility of higher loaded citrate_Th samples might be ascribed to the decreased amount of Fe-OH structure units and the increased amount of Fe-O-Si structure units. Hence, the increased amount of Fe-O-Fe structure units seemed not to determine the reducibility of higher loaded citrate_Th samples. Lower reducibility of higher loaded citrate_Th samples might facilitate formation of electrophilic surface oxygen species, yielding a higher selectivity towards total oxidation products at the expense of acrolein. For citrate_Th samples, highest acrolein selectivity was observed for lowest loaded sample. Lowest loaded citrate_Th sample predominantly consisted of dimeric Fe_xO_y and a low amount of isolated Fe^{3+} species. The high dispersion of iron oxidic species and the lower amount of Fe-O-Si bonds at lowest iron loading correlated with the lowest degree of micropore filling. Additionally, a high amount of Fe-OH structure units seemed likely. This presumably induced the highest reducibility of lowest loaded citrate_Th sample. Apparently, presence of dimeric and small oligomeric Fe_xO_y, and therefore, of reducible Fe-O-Fe structure units (Figure 6.27, (C)) yielded an optimal electron transfer efficiency during selective oxidation of propene. Furthermore, formation of nucleophilic surface oxygen species was presumably facilitated. This might explain the highest acrolein selectivity for lowest loaded citrate_Th sample.

For nitrate_Th samples, increasing iron loading correlated with an increasing degree of aggregation, and hence, an increasing number of both Fe-O-Si bonds (Figure 6.27, (D)) and Fe-O-Fe structure units (Figure 6.27, (C)). Lowest loaded nitrate_Th sample possessed the lowest degree of micropore filling due to the lowest amount of Fe-O-Si bonds and the highest dispersion of Fe_xO_y species. Compared to higher loaded nitrate_Th samples, reducibility of lowest loaded nitrate_Th sample was lower. This was ascribed to the higher dispersion of Fe_xO_y species and the lower amount of Fe-O-Fe structure units. Hence, formation of electrophilic surface oxygen species might be facilitated, yielding a higher selectivity towards total oxidation products at the expense of acrolein selectivity. Nitrate_Th samples with iron loadings higher than 2.0wt% consisted mainly of Fe_xO_y oligomers and α-Fe_2O_3 particles. Accordingly [FeO_6] units were not only edge-shared but also corner-shared (Figure 6.27, (E)). This induced a higher amount of Fe-O-Fe structure units, and hence, a better reducibility. Consequently, formation of nucleophilic O^{2-} surface oxygen species was presumably favored yielding a higher acrolein selectivity. However, highest loaded nitrate_Th sample possessed a significant amount of crystalline α-Fe_2O_3 which might hinder the accessibility of active iron sites for propene. Accordingly, acrolein selectivity decreased, despite the high reducibility and electron transfer efficiency.

The influence of layer thickness during calcination on Fe_xO_y/SBA-15 catalysts and their catalytic performance was dependent on the precursor. For all nitrate_Th samples, acrolein selectivity increased compared to corresponding nitrate samples. For lowest loaded nitrate_Th sample, increased acrolein selectivity correlated with a lower amount of isolated Fe^{3+} species, while that of Fe-O-Fe structure units increased. For iron loadings higher than 2.0wt%, increased acrolein selectivity of nitrate_Th compared to corresponding nitrate samples might be ascribed to the increased number of Fe-O-Fe structure units, and consequently to a higher reducibility and electron transfer efficiency. Formation of nucleophilic surface oxygen species was presumably facilitated in higher aggregated Fe_xO_y oligomers. Conversely, for citrate_Th samples, acrolein selectivity only increased for lowest loaded sample compared to corresponding citrate sample. This might be ascribed to the increased amount of Fe-O-Fe structure units at the expense of isolated Fe^{3+} species. The decreased acrolein selectivity of higher loaded citrate_Th compared to corresponding citrate samples might be ascribed to the higher degree of micropore filling, and hence, to the higher amount of Fe-O-Si bonds. Consequently, formation of electrophilic surface oxygen species was presumably facilitated. This might explain the decreased acrolein selectivity of citrate_Th samples with iron loadings higher than 2.5wt% compared to corresponding citrate samples.

6.7   Conclusion

In order to investigate the influence of powder layer thickness during calcination on Fe_xO_y/SBA-15 catalysts, calcination during synthesis of supported iron oxidic species on SBA-15 was conducted either in thin or thick powder layer. Fe_xO_y/SBA-15_Th catalysts consisted of small and well-dispersed iron oxidic species in the pore system of SBA-15. Structure of support material remained unaffected during synthesis. Fe_xO_y/SBA-15_Th catalysts were characterized with a multitude of analyzing methods.

Thick powder layer calcination extended the retention time of gaseous decomposition products of the precursor. Consequently, higher concentrations of gaseous decomposition products in powder layer during calcination enhanced iron oxidic species growth. This yielded higher aggregated and larger iron oxidic species compared to thin powder layer calcination. Higher aggregated and larger iron oxidic species of Fe_xO_y/SBA-15_Th compared to corresponding Fe_xO_y/SBA-15 catalysts for iron loadings higher than 2.5wt% correlated with an increased degree of micropore filling and a smoother SBA-15 surface. Iron oxidic species of citrate_Th samples were smaller and higher dispersed compared to those of nitrate_Th samples. This coincided with a higher degree of micropore filling and a smoother SBA-15 surface for citrate_Th compared to corresponding nitrate_Th samples for iron loadings higher than 2.5wt%. Hence, influence of precursor-dependent chelating effect on surface and porosity characteristics of Fe_xO_y/SBA-15 catalysts was observed independent of layer thickness during calcination. Citrate_Th samples possessed no long-range ordered phases. Conversely, for iron loadings higher than 2.0wt%, nitrate_Th samples consisted of a small fraction of α-Fe_2O_3 particles. Accordingly, precursor-dependent and iron loading-dependent particle size effects, i.e. larger iron oxidic species obtained from nitrate precursor and increasing species size with increasing iron loading, were also observed for Fe_xO_y/SBA-15_Th catalysts.

Besides structural characteristics, reducibility and solid-state kinetic properties of iron oxidic species supported on SBA-15 were affected by layer thickness during calcination. Differences in solid-state kinetic properties of Fe_xO_y/SBA-15 catalysts dependent on layer thickness during calcination correlated with differences in iron oxidic species size. TPR traces of citrate samples remained unaffected by layer thickness during calcination, despite a slight broadening of the first reduction peak for 2.5wt% Fe_Citrate_Th. Conversely, those of nitrate samples differed dependent on layer thickness during calcination. For lowest loaded nitrate sample, thick layer calcination induced a change from a single-step reduction mechanism to a multi-step reduction mechanism. Furthermore, thick layer calcination affected less resolved reduction peaks in the multi-step reduction mechanism for all nitrate samples. Lowest loaded Fe_xO_y/SBA-15_Th samples, 2.0wt% Fe_Nitrate_Th and 2.5wt% Fe_Citrate_Th, possessed a significantly increased apparent activation energy in reduction, determined by Kissinger method, compared to corresponding Fe_xO_y/SBA-15 samples. This coincided with the most pronounced increase in iron oxidic species size for these samples. As observed for nitrate samples, nitrate_Th samples showed a decreasing apparent activation energy in reduction with increasing iron loading. For iron loadings higher than 2.0wt%, nitrate_Th samples, consisting of largest and highest aggregated iron oxidic species, possessed an apparent activation energy in reduction similar to that of the mechanical mixture. Higher amount of α-Fe_2O_3 in these samples coincided with the increased reaction order of suitable solid-state reaction model and the increased topological dimension in reduction compared to all other Fe_xO_y/SBA-15_Th samples. Conversely, citrate_Th samples showed only minor differences in solid-state kinetic properties induced by thick layer calcination. For citrate samples, thick layer calcination did not lead to a change in suitable solid-state reaction model. Additionally, geometrical characteristics of the reduction remained unchanged. Apparent activation energy in reduction determined by Kissinger method showed a similar influence of the precursor as observed for Fe_xO_y/SBA-15 samples. For citrate_Th samples with iron loadings higher than 2.5wt%, higher values of apparent activation energy in reduction, compared to corresponding nitrate_Th samples, were observed. This was ascribed to the precursor-dependent chelating effect.

Layer thickness during calcination also affected catalytic performance in selective oxidation of propene. Reaction rates of propene conversion of Fe_xO_y/SBA-15_Th catalysts obtained from citrate precursor were higher compared to those obtained from nitrate precursor. This was ascribed to the precursor-dependent particle size effect. For citrate samples, differences in catalytic performance dependent on layer thickness during calcination correlated with differences in strengths of interactions between iron oxidic species and SBA-15. Increased acrolein selectivity of 2.5wt% Fe_Citrate_Th compared to 2.5wt% Fe_Citrate was ascribed to both weaker interactions between iron oxidic species and SBA-15 and less isolated Fe^{3+} species. Decreased acrolein selectivity of higher loaded citrate_Th samples correlated with stronger interactions between iron oxidic species and SBA-15. Accordingly, formation of electrophilic surface oxygen was presumably facilitated. Conversely, for all nitrate_Th samples, an increased acrolein selectivity compared to nitrate samples was observed. Increased acrolein selectivity of 2.0wt% Fe_Nitrate_Th was ascribed to weaker interactions between iron oxidic species and SBA-15 and less isolated Fe^{3+} species. Increased acrolein selectivity of 7.2wt% Fe_Nitrate_Th and 9.3wt% Fe_Nitrate_Th, compared to corresponding nitrate samples, corelated with a decreased apparent activation energy in reduction. Furthermore, higher aggregated Fe_xO_y species might induce a better electron transfer efficiency and a facilitated formation of nucleophilic surface oxygen.

Detailed characterization of Fe_xO_y/SBA-15_Th and Fe_xO_y/SBA-15 catalysts identified powder layer thickness during calcination as further synthesis parameter with a significant influence on resulting iron oxidic species supported on SBA-15. Layer thickness during calcination of Fe_xO_y/SBA-15 catalysts influenced not only their structural characteristics, but also their reducibility, solid-state kinetic properties, and catalytic performance in selective oxidation of propene. The magnitude of influence of layer thickness during calcination on Fe_xO_y/SBA-15 catalysts was affected by both iron loading and Fe(III) precursor.

7 Influence of molybdenum addition on Fe_xO_y/SBA-15 catalysts

7.1 Introduction

Deducing structure-activity correlations of model catalysts constitutes a major field of catalysis research. Therefore, a detailed knowledge of the catalysts structure is indispensable. In the previous chapters, a detailed characterization of the Fe_xO_y/SBA-15 catalysts was presented. Based on these results, in this chapter, it will be shown that molybdenum addition to Fe_xO_y/SBA-15 catalysts has a significant influence on iron oxide dispersion, reducibility, and catalytic performance. $Mo_xO_y\_Fe_xO_y$/SBA-15 catalysts were investigated by various characterization methods, such as N_2 physisorption, XRD, and DR-UV-Vis spectroscopy. Additionally, reducibility and catalytic performance of $Mo_xO_y\_Fe_xO_y$/SBA-15 catalysts in selective oxidation of propene were studied.

7.2 Experimental

7.2.1 Sample preparation

Mesoporous silica SBA-15 was prepared according to Zhao et al. [33, 34] as described in chapter 4.2.1. Mixed iron and molybdenum oxide catalysts supported on SBA-15 were prepared by incipient wetness technique and denoted $Mo_xO_y\_Fe_xO_y$/SBA-15. Therefore, an aqueous solution of ammonium iron(III) citrate ($\sim$ 18% Fe, Roth) and ammonium heptamolybdate tetrahydrate ($\geq$ 99%, Fluka) was used. The pH value of the aqueous oxide precursor solution was adjusted to 7.5-8. After drying in air for 24 h, calcination was carried out at 723 K for 5 h. To investigate the influence of molybdenum addition on Fe_xO_y/SBA-15 catalysts, iron loading was kept invariant, while Mo/Fe atomic ratio was varied between 0.07/1.0 and 0.57/1.0. According to the Mo/Fe atomic ratio, samples were denoted 0.07Mo/1.0Fe/SBA-15, 0.10Mo/1.0Fe/SBA-15, 0.15Mo/1.0Fe/SBA-15, 0.21Mo/1.0Fe/SBA-15, and 0.57Mo/1.0Fe/SBA-15. Furthermore, a sample with 1.3wt% Mo supported on SBA-15, 1.3wt% Mo_SBA-15, was prepared as reference. Metal oxide loadings of the samples were quantified by X-ray fluorescence spectroscopy. Moreover, CHN analysis was performed to confirm a complete decomposition of the precursors.

Independent of the Mo/Fe atomic ratio, uncalcined $Mo_xO_y\_Fe_xO_y$/SBA-15 samples were light yellow and turned orange brownish after calcination. In contrast, sample 1.3wt% Mo_SBA-15 was colorless before and after calcination, while sample 10.7wt% Fe_Citrate was also light yellow before and orange brownish after calcination. Figure 7.1 illustrates the change in color of the samples.

Uncalcined **Calcined**

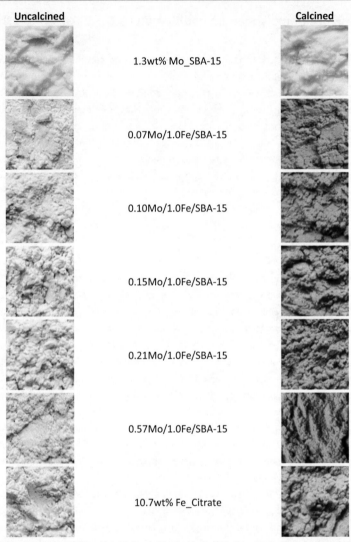

1.3wt% Mo_SBA-15

0.07Mo/1.0Fe/SBA-15

0.10Mo/1.0Fe/SBA-15

0.15Mo/1.0Fe/SBA-15

0.21Mo/1.0Fe/SBA-15

0.57Mo/1.0Fe/SBA-15

10.7wt% Fe_Citrate

Figure 7.1: 1.3wt% Mo_SBA-15 (top), $Mo_xO_y\_Fe_xO_y$/SBA-15 samples (increasing Mo/Fe atomic ratio from top to bottom), and 10.7wt% Fe_Citrate (bottom) before (left) and after (right) calcination.

7.2.2 Sample characterization

X-ray fluorescence analysis

Quantitative analysis of the metal oxide loadings on SBA-15 was conducted by X-ray fluorescence spectroscopy on an AXIOS X-ray spectrometer (2.4 kW model, PANalytical), equipped with a Rh K_α X-ray source, a gas flow detector, and a scintillation detector. Prior to measurements, samples were mixed with wax (Hoechst, Merck), ratio 1:1, and pressed into pellets of 13 mm diameter. Quantification was performed by standardless analysis using the software package SuperQ5 (PANalytical).

CHN elemental analysis

Elemental contents of C, H, and N were determined using a Thermo FlashEA 1112 Organic Elemental Analyzer (ThermoFisher Scientific) with CHNS-O configuration. Measurements were conducted in collaboration with the measuring center at institute of chemistry at TU Berlin.

Powder X-ray diffraction

Powder X-ray diffraction patterns were obtained using an X'Pert PRO diffractometer (PANalytical, 40 kV, 40 mA) in theta/theta geometry equipped with a solid-state multi-channel detector (PIXel). Cu K_α radiation was used. Wide-angle diffraction scans were conducted in reflection mode. Small-angle diffraction patterns were measured in transmission mode from 0.4° through 6° 2θ in steps of 0.013° 2θ with a sampling time of 90 s/step.

Nitrogen physisorption

Nitrogen adsorption/desorption isotherms were measured at 77 K using a BELSORP Mini II (BEL Inc. Japan). Prior to measurements, the samples were pre-treated under reduced pressure (10^{-2} kPa) at 368 K for 35 min and kept under the same pressure at 448 K for 15 h (BELPREP II vac).

DR-UV-Vis spectroscopy

Diffuse reflectance UV-Vis (DR-UV-Vis) spectroscopy was conducted on a two-beam spectrometer (V-670, Jasco) using a barium sulfate coated integration sphere. (Scan speed 100 nm/min, slit width 5.0 nm (UV-Vis) and 20.0 nm (NIR), and spectral region 220-2000 nm). SBA-15 was used as white standard for all samples.

Temperature-programmed reduction

Temperature-programmed reduction (TPR) was performed using a BELCAT_B (BEL Inc. Japan). Samples were placed on silica wool in a silica glass tube reactor. Evolving water was trapped using a molecular sieve (4 Å). Gas mixture consisted of 5% H_2 in 95% Ar with a total gas flow of 40 ml/min. Heating rates used were 5, 10, 15, and 20 K/min to 1223 K. A constant initial sample weight of 0.03 g was used and H_2 consumption was continuously monitored by a thermal conductivity detector.

Catalytic measurements

Quantification of catalytic activity in selective oxidation of propene was performed using a laboratory fixed-bed reactor connected to an online gas chromatography system (CP-3800, Varian) and a mass spectrometer (Omnistar, Pfeiffer Vacuum). The fixed-bed reactor consisted of a SiO_2 tube (30 cm length, 9 mm inner diameter) placed vertically in a tube furnace. A P3 frit was centered in the SiO_2 tube in the isothermal zone, where the sample was placed. The samples were diluted with boron nitride (99.5%, Alfa Aesar) to achieve a constant volume in the reactor and to minimize thermal effects. Overall sample masses were 0.25 g. The reactor was operated at low propene conversion levels (5-10%) to ensure differential reaction conditions. For catalytic testing in selective oxidation of propene, a gas mixture of 5% propene and 5% oxygen in helium was used in a temperature range of 298-653 K with a heating rate of 5 K/min. Reactant gas flow rates of propene, oxygen, and helium were adjusted through separate mass flow controllers (Bronkhorst) to a total flow of 40 ml/min. All gas lines and valves were preheated to 473 K. Hydrocarbons and oxygenated reaction products were analyzed using a Carbowax 52CB capillary column, connected to an Al_2O_3/MAPD capillary column (25 m x 0.32 mm) or a fused silica restriction (25 m x 0.32 mm), each connected to a flame ionization detector (FID). Permanent gases (CO, CO_2, N_2, O_2) were separated and analyzed using a "Permanent Gas Analyzer" (CP-3800, Varian) with a Hayesep Q (2 m x 1/8″) and a Hayesep T packed column (0.5 m x 1/8″) as precolumns combined with a back flush. For separation, a Hayesep Q packed column (0.5 m x 1/8″) was connected via a molecular sieve (1.5 m x 1/8″) to a thermal conductivity detector (TCD). Additionally, product and reactant gas flow were continuously monitored by a connected mass spectrometer (Omnistar, Pfeiffer) in a multiple ion detection mode. Details on calculation of catalytic parameters, such as conversion, selectivity, reaction rate, and carbon balance, are described in chapter 5.2.2.

7.3 Structural characterization of $Mo_xO_y\_Fe_xO_y$/SBA-15 catalysts

7.3.1 Mesoporous structure of $Mo_xO_y\_Fe_xO_y$/SBA-15

Long-range ordered structure

Long-range ordered structure of the $Mo_xO_y\_Fe_xO_y$/SBA-15 samples was investigated by wide-angle and small-angle X-ray diffraction. Figure 7.2, left depicts wide-angle X-ray diffraction patterns of all $Mo_xO_y\_Fe_xO_y$/SBA-15 samples. The absence of sharp diffraction peaks was indicative of no long-range ordered phases and high dispersion of small iron and molybdenum oxides within the $Mo_xO_y\_Fe_xO_y$/SBA-15 samples. The broad diffraction peak at 23° 2θ corresponded to the amorphous SBA-15 support material. [49] Small-angle diffraction patterns of all samples exhibited characteristic diffraction peaks (10l), (11l), and (20l) of SBA-15 (Figure 7.2, right). [32, 33, 101] Hence, two-dimensional hexagonal symmetry of the support material SBA-15 was retained after supporting iron and molybdenum oxides.

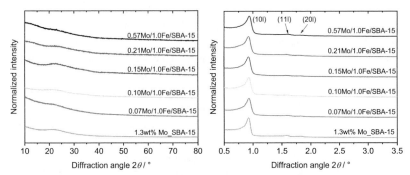

Figure 7.2: Left: Wide-angle X-ray diffraction patterns of $Mo_xO_y\_Fe_xO_y$/SBA-15 samples at various Mo/Fe atomic ratios and sample 1.3wt% Mo_SBA-15. Right: Small-angle X-ray diffraction patterns of $Mo_xO_y\_Fe_xO_y$/SBA-15 samples and sample 1.3wt% Mo_SBA-15.

Surface and porosity characteristics

N_2 physisorption measurements were conducted for determining specific surface area and investigating pore structure of the $Mo_xO_y\_Fe_xO_y$/SBA-15 samples. All samples exhibited type IV nitrogen adsorption/desorption isotherms with H1 hysteresis loops indicating mesoporous materials (Figure 7.3). [68] Adsorption and desorption branches were nearly parallel at the hysteresis loop, as expected for regularly shaped pores. However, a knee in the hysteresis loop was observed for all $Mo_xO_y\_Fe_xO_y$/SBA-15 samples except for sample 1.3wt% Mo_SBA-15. A knee in the hysteresis loop might have three possible reasons. First, synthesis procedure could have varied the structure of SBA-15. A changed SBA-15 structure was excluded based on XRD results confirming structure preservation. Furthermore, despite the observed knee in the hysteresis loop, N_2 adsorption/desorption isotherms of all samples were identified to be type IV with H1 hysteresis loop, being characteristic for SBA-15 support material. Second, partially blocked mesopores of SBA-15 could have yielded a knee in the hysteresis loop. This possible reason was also excluded since desorption and adsorption branches converged after the hysteresis loop at higher p/p_0, and all samples possessed similar pore radius distributions (Figure 7.4). Besides, partial blocking of mesopores, as reported for CuNi/SBA-15 (5wt% Cu, 5wt% Ni) [174], NiO/SBA-15 (24wt% Ni) [175], or CoMoW/SBA-15 (9wt% MoO_3, 14wt% WO_3, 4wt% CoO) [176], was always associated with pronounced two-step desorption branches, whereas adsorption branches were unaffected. Hence, the third possible reason, a bimodal particle size distribution, seemed to be reasonable. The knee in the hysteresis loop of the $Mo_xO_y\_Fe_xO_y$/SBA-15 samples, was probably induced by bimodal particle size distributions, complying with pore radius distributions determined by BJH method. Samples 0.07Mo/1.0Fe/SBA-15, 0.10Mo/1.0Fe/SBA-15, 0.15Mo/1.0Fe/SBA-15, and 0.21Mo/1.0Fe/SBA-15 showed a bimodal pore radius distribution with a main maximum at 4 nm and a second maximum at $\sim$ 3-3.5 nm. In contrast, pore radius distribution of 1.3wt% Mo_SBA-15 was narrow with a sharp maximum at 4 nm and that of 0.57Mo/1.0Fe/SBA-15 was broad and ranged from $\sim$ 2.6 through $\sim$ 4.6 nm (Figure 7.4). Differences in pore radius contributions dependent on Mo/Fe atomic ratio probably originated from various interactions between molybdenum and iron during synthesis. At low Mo/Fe atomic ratios, significantly higher amounts of iron compared to molybdenum were present in the $Mo_xO_y\_Fe_xO_y$/SBA-15 samples. Hence, only a small fraction of iron atoms directly interacted with molybdenum atoms during synthesis, resulting in a bimodal particle size distribution. Fe_xO_y species which were directly influenced by molybdenum atoms differed in species size compared to Fe_xO_y without direct influence of molybdenum atoms. Accordingly, differences in chemical environment during synthesis were probably responsible for bimodal particle size distributions, and hence, for the knee in the hysteresis loop of N_2 adsorption/desorption isotherms of the $Mo_xO_y\_Fe_xO_y$/SBA-15 samples. For sample 0.57Mo/1.0Fe/SBA-15, possessing a more balanced Mo/Fe atomic ratio, such differences in chemical environment were less pronounced. The higher Mo/Fe atomic ratio effected more balanced interactions between iron and molybdenum atoms during synthesis, inducing a broadened particle size distribution. However, the broadened particle size distribution agreed with the knee in the hysteresis loop because of the still existing excess of iron compared to molybdenum.

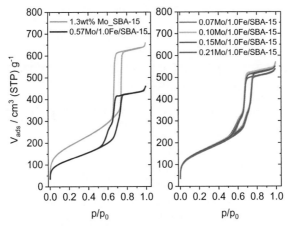

Figure 7.3: N$_2$ adsorption/desorption isotherms of Mo$_x$O$_y$_Fe$_x$O$_y$/SBA-15 samples and sample 1.3wt% Mo_SBA-15.

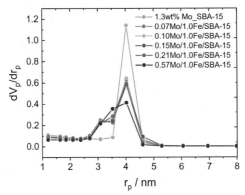

Figure 7.4: Pore radius distribution of Mo$_x$O$_y$_Fe$_x$O$_y$/SBA-15 samples and 1.3wt% Mo_SBA-15.

The specific surface area, $a_{s,BET}$, of Mo$_x$O$_y$_Fe$_x$O$_y$/SBA-15 samples systematically decreased with increasing Mo/Fe atomic ratio. Whereas SBA-15 possessed a specific surface area between 706.8 and 756.8 m^2/g, those after supporting iron and molybdenum oxides were determined to be between 447.6 and 675.7 m^2/g. This decrease in $a_{s,BET}$ together with a decrease in pore volume, V_{pore}, from 1.036-1.203 cm^3/g of SBA-15 to 0.705-1.012 cm^3/g of as-prepared samples, indicated the iron and molybdenum species being successfully located in the mesopores of SBA-15. Details on surface and porosity characteristics evaluated by N$_2$ physisorption and XRD measurements are summarized in Table 7.1.

Table 7.1: Surface and porosity characteristics of $Mo_xO_y\_Fe_xO_y$/SBA-15 samples, sample 1.3wt% Mo_SBA-15, and sample 10.7wt% Fe_Citrate. Specific surface area, $a_{s,BET}$, pore volume, V_{pore}, ratio of mesopore surface area, a_{pore}, and specific surface area, $a_{s,BET}$, as measure of micropore contribution to the entire surface of SBA-15, lattice constant, a_0, of hexagonal unit cell, wall thickness between the mesopores of SBA-15, d_w, and differences in fractal dimension, ΔD_f, between SBA-15 and corresponding $Mo_xO_y\_Fe_xO_y$/SBA-15 samples as measure of the roughness of the surface.

	$a_{s,BET}$ / m^2/g	V_{pore} / cm^3/g	$a_{pore}/a_{s,BET}$	a_0 / nm	d_w / nm	ΔD_f
1.3wt% Mo_SBA-15	675.7 ± 0.7	1.012 ± 0.001	0.88	11.14 ± 0.01	3.08 ± 0.01	0.06 ± 0.02
0.07Mo/1.0Fe/SBA-15	581.1 ± 0.6	0.865 ± 0.001	0.91	11.09 ± 0.01	3.03 ± 0.01	0.11± 0.02
0.10Mo/1.0Fe/SBA-15	579.1 ± 0.6	0.869 ± 0.001	0.91	11.01 ± 0.01	2.95 ± 0.01	0.13± 0.01
0.15Mo/1.0Fe/SBA-15	573.2 ± 0.6	0.851 ± 0.001	0.90	11.17 ± 0.01	3.11 ± 0.01	0.11 ± 0.02
0.21Mo/1.0Fe/SBA-15	554.7 ± 0.7	0.832 ± 0.001	0.91	11.18 ± 0.01	3.12 ± 0.01	0.14 ± 0.02
0.57Mo/1.0Fe/SBA-15	447.6 ± 0.4	0.705 ± 0.001	0.96	11.05 ± 0.01	2.99 ± 0.01	0.27± 0.03
10.7wt% Fe_Citrate	605.7 ± 0.6	0.898 ± 0.001	0.91	10.94 ± 0.01	2.88 ± 0.01	0.15 ± 0.02

Contribution of micropores to entire surface area

Mesopore surface area, a_{pore}, calculated by BJH method [69] and specific surface area, $a_{s,BET}$, calculated by BET method [70] can be used for estimating the contribution of micropores to the entire surface area of SBA-15. [71, 110] Therefore, the ratio of mesopore surface area and specific surface area, $a_{pore}/a_{s,BET}$, was calculated (Table 7.1). For $Mo_xO_y\_Fe_xO_y$/SBA-15 samples with atomic ratios between 0.07Mo/1.0Fe and 0.21Mo/1.0Fe, ratio of $a_{pore}/a_{s,BET}$ was nearly invariant at 0.91 and 0.90, respectively. Ratio of $a_{pore}/a_{s,BET}$ of 0.91 was also obtained for sample 10.7wt% Fe_Citrate, the Fe_xO_y/SBA-15 sample without molybdenum addition. Hence, molybdenum addition at atomic ratios between 0.07Mo/1.0Fe and 0.21Mo/1.0Fe did not affect the contribution of micropores to the entire surface area of SBA-15. In contrast, a higher atomic ratio of 0.57Mo/1.0Fe induced an increased ratio of $a_{pore}/a_{s,BET}$ up to 0.96 indicating a decreased contribution of micropores. Conversely, an increased contribution of micropores to the entire surface area of SBA-15 was observed for sample 1.3wt% Mo_SBA-15 possessing a ratio of $a_{pore}/a_{s,BET}$ of 0.88. The low molybdenum loading of this sample correlated with small and highly dispersed molybdenum species. Therefore, a lower degree of micropore filling was observed.

Surface roughness

In addition to BET and BJH method, modified FHH method was used to analyze the nitrogen physisorption data. Herein, the fractal dimension D_f was determined as a measure of the roughness of the surface. [71, 72] For $Mo_xO_y\_Fe_xO_y$/SBA-15 samples, 1.3wt% Mo_SBA-15, 10.7wt% Fe_Citrate, and SBA-15, the fractal dimension ranged between 2 and 3, indicating a rough surface. [86] In order to elucidate the effect of supported iron and molybdenum species on surface roughness of the support material, ΔD_f values were calculated as difference between D_f values of SBA-15 and those of corresponding $Mo_xO_y\_Fe_xO_y$/SBA-15 samples. For $Mo_xO_y\_Fe_xO_y$/SBA-15 samples with atomic ratios between 0.07Mo/1.0Fe and 0.21Mo/1.0Fe, ΔD_f values were invariant within the error limits. Consequently, molybdenum addition at low Mo/Fe atomic ratios did not influence the surface roughness of SBA-15. An invariant surface roughness of support material at atomic ratios between 0.07Mo/1.0Fe and 0.21Mo/Fe coincided with a nearly invariant contribution of micropores to the entire surface area of SBA-15, $a_{pore}/a_{s,BET}$.

Only for samples 1.3wt% Mo_SBA-15 and 0.57Mo/1.0Fe/SBA-15, ΔD_f was significantly lower or higher, respectively. For sample 1.3wt% Mo_SBA-15, consisting of small and highly dispersed Mo_xO_y species, lowest value of ΔD_f agreed with the highest contribution of micropores to the entire surface area of SBA-15. At this low metal oxide loading, most of the pores of SBA-15 remained unfilled, and hence, surface roughness of SBA-15 was only slightly affected. The higher ΔD_f value for sample 0.57Mo/1.0Fe/SBA-15 indicated a smoother surface of SBA-15. The decreased contribution of micropores to the entire surface area of SBA-15 for this sample was in accordance with the smoother surface. Deposition of iron and molybdenum oxides in the pores of SBA-15 at higher Mo/Fe atomic ratio resulted in an enhanced micropore filling, accompanied by an enhanced smoothing of the support surface.

7.3.2 Diffuse reflectance UV-Vis spectroscopy

Chemical environment of the supported iron and molybdenum centers was investigated by DR-UV-Vis spectroscopy. DR-UV-Vis spectra of $Mo_xO_y\_Fe_xO_y$/SBA-15 samples showed a broad absorption, resulting from overlapping absorption bands, while absorption of sample 1.3wt% Mo_SBA-15 was significantly narrower. Figure 7.5 depicts DR-UV-Vis spectra of all $Mo_xO_y\_Fe_xO_y$/SBA-15 samples and 1.3wt%Mo_SBA-15 before and after calcination.

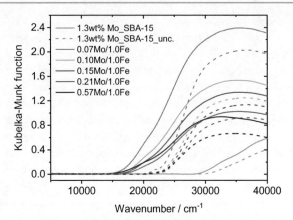

Figure 7.5: DR-UV-Vis spectra of $Mo_xO_y\_Fe_xO_y$/SBA-15 samples at Mo/Fe atomic ratios between 0.07/1.0 and 0.57/1.0 and 1.3wt% Mo_SBA-15 before (dashed lines) and after (straight lines) calcination.

DR-UV-Vis spectra of $Mo_xO_y\_Fe_xO_y$/SBA-15 samples exhibited a similar shape before and after calcination. However, calcination of the samples induced a red-shift of the absorption and an increased value of Kubelka-Munk function, independent of Mo/Fe atomic ratio. Increasing Mo/Fe atomic ratio correlated with a decreasing value of Kubelka-Munk function.

The average metal oxide particle size was estimated by analyzing the DR-UV-Vis edge energy, determined from the position of the low-energy rise in DR-UV-Vis spectra. Analogous to the model of the particle in the box, decreasing DR-UV-Vis edge energy is reported to correlate with an increasing size of transition metal oxide domain. [49, 114, 115] DR-UV-Vis edge energies of $Mo_xO_y\_Fe_xO_y$/SBA-15 samples, 1.3wt% Mo_SBA-15, and iron oxide references were determined according to Weber [114], and depicted as function of Mo/Fe atomic ratio (Figure 7.6). Sample 1.3wt% Mo_SBA-15 possessed the highest value of DR-UV-Vis edge energy, and hence, the smallest Mo_xO_y species size. Determined edge energy of 4.24 eV was close to that reported for mononuclear molybdate species, $[MoO_4]^{2-}$, indicating that this sample mainly consisted of isolated mononuclear Mo_xO_y species. [114, 177] The smallest species size together with the presence of mainly isolated, mononuclear species was in accordance with the low metal oxide loading and the results from N_2 physisorption measurements. All $Mo_xO_y\_Fe_xO_y$/SBA-15 samples possessed edge energy values higher than those of crystalline α-Fe_2O_3 and 10.7wt% Fe_Citrate, corresponding Fe_xO_y/SBA-15 catalyst without molybdenum addition. Accordingly, molybdenum addition induced an increased iron oxide dispersion, and furthermore, a decreased average species size. A dispersion effect of molybdenum on iron oxidic species was also reported for Mo-Fe catalysts supported on activated carbon [58] and Fe-Mo catalysts supported on Al_2O_3 [59]. DR-UV-Vis edge energy of the $Mo_xO_y\_Fe_xO_y$/SBA-15 samples increased with increasing Mo/Fe atomic ratio between 0.07/1.0 and 0.21/1.0, indicating a decreasing average species size. Apparently, dispersion effect of molybdenum on Fe_xO_y species was enhanced by increasing Mo/Fe atomic ratio up to 0.21/1.0. A further increased Mo/Fe atomic ratio to 0.57/1.0 yielded a slightly decreased DR-UV-Vis edge energy, indicating a slightly increased average species size. It seemed reasonable that Mo/Fe atomic ratio of 0.57/1.0 induced a slightly increased Mo_xO_y species size due to the increased molybdenum loading. However, Fe_xO_y species remained still smaller and higher dispersed on the support compared to those without molybdenum addition.

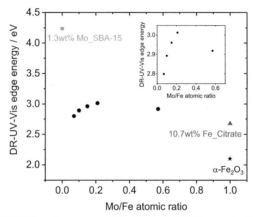

Figure 7.6: DR-UV-Vis edge energy of Mo_xO_y_Fe_xO_y/SBA-15 samples (dots), 1.3wt% Mo_SBA-15 (square), 10.7wt% Fe_Citrate (triangle), and α-Fe_2O_3 (star) as function of Mo/Fe atomic ratio. Inset depicts the enlarged trend for the Mo_xO_y_Fe_xO_y/SBA-15 samples.

To further elucidate the influence of molybdenum addition on chemical environment of iron centers on SBA-15, DR-UV-Vis spectra of the calcined Mo_xO_y_Fe_xO_y/SBA-15 samples were fitted with Gaussian functions. Full width at half maximum (FWHM) of the fitted Gaussian functions were correlated to be equal. Additionally, F-Tests were conducted to ensure statistically meaningful results. DR-UV-Vis spectra of Mo_xO_y_Fe_xO_y/SBA-15 samples and 10.7wt% Fe_Citrate are depicted in Figure 7.7, together with refined profile functions. For all spectra, three Gaussian functions were necessary to describe the broad absorption band. Fitting procedure was performed analogously to that described in chapter 5.3.3.

Commonly, characteristic absorption bands in high energy range, at wavenumbers above 33333 cm^{-1}, are attributed to charge transfer transitions (CT) of isolated Fe^{3+} and isolated Mo^{6+}. [48, 57, 112] Charge transfer transitions of octahedrally coordinated Fe^{3+} in small dimeric or oligomeric Fe_xO_y are expected in the wavenumber range from 25000 through 33333 cm^{-1}. Absorption bands below 25000 cm^{-1} are mainly assigned to charge transfer transitions in α-Fe_2O_3 nanoparticles. [48, 50, 112, 117, 178] Charge transfer transitions in iron oxidic species usually result in intense and broad absorption bands. The wavenumber of these absorption bands is dependent on both coordination number of oxygen ligands and degree of aggregation. Conversely, symmetry and spin forbidden single electron d-d transitions in Fe^{3+} ions result in weak absorption bands at wavenumbers below 20000 cm^{-1}. [112, 113]

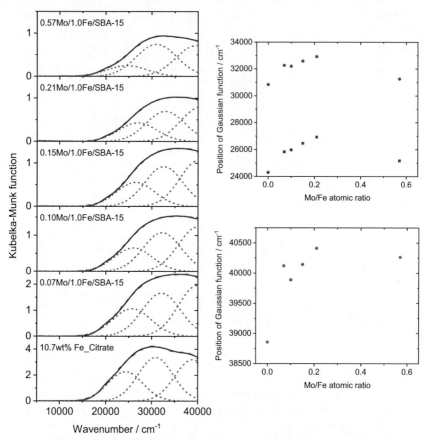

Figure 7.7: Left: Refinement of the sum function (dashed lines) of the fitted Gaussian functions (dotted lines) to experimental DR-UV-Vis spectra (straight lines). Right: Positions of Gaussian functions in DR-UV-Vis spectra as function of Mo/Fe atomic ratio.

Neither sample 10.7wt% Fe_Citrate nor any $Mo_xO_y\_Fe_xO_y$/SBA-15 sample possessed absorption bands below 20000 cm^{-1} attributed to d-d transitions. This was indicative of a symmetric coordination of oxygen atoms around the iron centers in supported iron species. Without molybdenum addition, the first absorption band, at 24306 cm^{-1}, was correlated with the presence of small α-Fe_2O_3 nanoparticles. However, corresponding first absorption band of the $Mo_xO_y\_Fe_xO_y$/SBA-15 samples showed a blue-shift towards wavenumbers higher than 25000 cm^{-1}. This blue-shift was indicative of the dispersion effect of molybdenum on Fe_xO_y species. Molybdenum addition induced a higher dispersion of Fe_xO_y species combined with a hindered aggregation, and thus, resulted in absorption bands of CT transitions in dimeric or small oligomeric Fe_xO_y instead of CT transitions in small α-Fe_2O_3 nanoparticles. The dispersion effect of molybdenum also affected the second absorption band. For sample 10.7wt% Fe_Citrate, this absorption band was located at 30839 cm^{-1} and ascribed to CT transitions of octahedrally coordinated Fe^{3+} in Fe_xO_y oligomers. Molybdenum addition induced a blue-shift of this absorption band. Commonly, a red-shift of the absorption band of charge transfer transitions of Fe^{3+} is correlated with an increasing number of coordinating oxygen ligands. Based on this correlation, the blue-shift of the second absorption band induced by molybdenum addition was ascribed to a decreased number of oxygen ligands, and further, to an increased dispersion of the iron oxidic species on SBA-15. The third absorption band was located at wavenumbers above 33333 cm^{-1} and attributed to CT transitions of O^{2-} ligands to isolated octahedrally coordinated Fe^{3+}. Again, molybdenum addition induced a blue-shift of the absorption band, indicative of an increased dispersion of the iron species. However, due the broad absorption band above 33333 cm^{-1}, distinction between CT transitions of O^{2-} to Fe^{3+} and O^{2-} to Mo^{6+} in octahedral or tetrahedral coordination sphere was not possible.

Figure 7.7 depicts the positions of the three Gaussian functions as function of Mo/Fe atomic ratio. The distinct blue-shift of all absorption bands due to the dispersion effect of molybdenum on Fe_xO_y species was visible for all Mo_xO_y_Fe_xO_y/SBA-15 samples compared to 10.7wt% Fe_Citrate. This blue-shift increased with increasing Mo/Fe atomic ratio up to 0.21/1.0. Further increasing Mo/Fe atomic ratio yielded a slight red-shift of the three absorption bands compared to the samples at lower Mo/Fe atomic ratios. Both an increased species size due to the highest metal oxide loading and a formation of less dispersed Mo_xO_y species might have induced this red-shift. Characteristic absorption bands of crystalline $Fe_2(MoO_4)_3$ were expected at 18789, 21739, and 30769 cm^{-1}. [179, 180] The absence of these bands agreed with the results from XRD and N_2 physisorption measurements, indicative of the presence of only small and highly dispersed species on the support material. However, contribution of charge transfer transitions in Fe-O-Mo units in small oligomeric species was not excluded.

In contrast to Mo_xO_y_Fe_xO_y/SBA-15 and 10.7wt% Fe_Citrate, DR-UV-Vis spectrum of 1.3wt% Mo_SBA-15 was fitted using only two Gaussian functions. Fitted Gaussian functions, together with the sum function and the experimental spectrum are depicted in Figure 7.8. Overlapping absorption bands in the wavenumber range between 33333 and 45455 cm^{-1} are reported to be assigned to charge transfer transitions in MoO_x species with both octahedral and tetrahedral coordination environment. Conversely, absorption bands between 38462 and 43478 cm^{-1} are commonly assigned to Mo=O of isolated tetrahedral units. [57, 179, 181] The two absorption bands of 1.3wt% Mo_SBA-15 located at 34375 and 39734 cm^{-1} were in accordance with the presence of small and isolated Mo_xO_y species on SBA-15 for this low loaded molybdenum sample. A combination of tetrahedral and octahedral environment of the Mo^{6+} centers seemed reasonable due to the pH dependence of the structure of Mo_xO_y species. (see chapter 7.3.3).

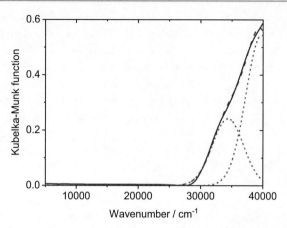

Figure 7.8: Refinement of the sum function (dashed lines) of the fitted Gaussian functions (dotted lines) to experimental DR-UV-Vis spectrum (straight lines) of 1.3wt% Mo_SBA-15.

7.3.3 Possible interactions between ferric and molybdic species

Iron and molybdenum interactions during synthesis

For synthesis of $Mo_xO_y\_Fe_xO_y$/SBA-15 catalysts, aqueous solutions of iron and molybdenum oxide precursors were used. Unlike iron oxide species, which are predominantly present as stable and molecular Fe(III) species in aqueous solutions, structure of molybdate species possesses a high pH dependence. [182] Dissolving ammonium heptamolybdate tetrahydrate in aqueous solutions at pH values higher than 6.5 results in isolated tetrahedral MoO_4^{2-} species, whereas lower pH values yield octahedrally coordinated polyoxoanions, such as heptamolybdate $Mo_7O_{24}^{6-}$ or octamolybdate $Mo_8O_{28}^{4-}$. [183–185] Therefore, pH value of aqueous precursor solutions had to be precisely adjusted to values higher than 6.5. Based on synthesis procedure (see chapter 7.2.1) at pH 7.5-8, Fe(III) ions in the dissolved ammonium iron(III) citrate precursor directly interacted with Mo(VI) ions in the dissolved ammonium heptamolybdate tetrahydrate precursor. Hence, formation of iron and molybdenum mixed oxides was preferred. However, the local structure of supported metal oxides is known to be strongly influenced by the nature of the support and by the hydrated and dehydrated state. The hydrated state of supported metal oxides resembles the oxide structure in aqueous solution at similar pH value. [30, 114, 183, 185] Hence, after applying the precursor solution on the acidic support material, interactions between Fe(III) and Mo(VI) ions might have been influenced by surface hydroxyl groups of SBA-15.

Interactions between iron and molybdenum in resulting Fe-O-Mo structure units

Several studies suggested that Fe_2O_3 and MoO_3 can easily form $Fe_2(MoO_4)_3$ after heat-treatment at temperatures higher than 400 °C. [58–60, 62] Ferric molybdate, $Fe_2(MoO_4)_3$, crystallizes in a monoclinic crystal system. The structure of this phase consists of isolated

[FeO_6] octahedra and [MoO_4] tetrahedra that share corners in which each oxygen atom is bound only to one iron and one molybdenum atom. [59, 60]

For supported iron and molybdenum oxide systems with metal oxide loadings of 12wt% Mo and 15.7wt% Fe, formation of crystalline ferric molybdate phase is reported to occur at Mo/Fe atomic ratios higher than 0.76/1.0. [58] These metal oxide loadings are higher compared to those of the $Mo_xO_y\_Fe_xO_y$/SBA-15 catalysts. However, the presence of amorphous highly dispersed ferric molybdate phase, as reported by Qin et al. [60] for low iron and molybdenum contents, seemed possible. The strong interactions between iron and molybdenum during synthesis might lead to formation of species with a local geometry similar to that of $Fe_2(MoO_4)_3$.

7.4 *In situ* characterization

7.4.1 Reducibility of $Mo_xO_y\_Fe_xO_y$/SBA-15 catalysts

Gaining insight into reducibility of $Mo_xO_y\_Fe_xO_y$/SBA-15 samples was an important starting point for elucidating the influence of molybdenum addition on catalytic performance of Fe_xO_y/SBA-15 catalysts for selective oxidation of propene. Selective oxidation reactions proceed according to the so-called Mars-van-Krevelen mechanism where catalysts are partial reduced and re-oxidized during catalytic cycle. [7, 12, 27] Suitable selective oxidation catalysts must possess an intermediate metal-oxygen bond strength. A too weak metal-oxygen bond strength will lead to mainly total oxidation products. Conversely, if the metal-oxygen bond strength is too strong, catalytic reaction will not proceed. [7]

Figure 7.9 depicts TPR traces of all $Mo_xO_y\_Fe_xO_y$/SBA-15 samples, 10.7wt% Fe_Citrate, and 1.3wt% Mo_SBA-15. Significant differences in TPR traces were discernible. 10.7wt% Fe_Citrate showed a two-step reduction mechanism where the first and second reduction step were assigned to reduction of Fe(III) oxidic species to Fe(II) oxidic species and further reduction to Fe(0), respectively. [86] Molybdenum addition at atomic ratios between 0.07Mo/1.0Fe and 0.21Mo/1.0Fe induced a rising shoulder at the first TPR maxima at higher temperatures while the TPR maxima were shifted to higher temperatures. Conversely, at an atomic ratio of 0.57Mo/1.0Fe, the first TPR maximum was sharp but further shifted to higher temperatures. The shift of the first TPR maxima towards higher temperatures (Figure 7.10) indicated a lower reducibility of the Fe_xO_y species. A decreased reducibility due to molybdenum addition to Fe_xO_y systems, and a further decreasing reducibility with increasing molybdenum content, was in accordance with the literature. [57, 59, 60] Apparently, molybdenum addition induced both an electronic and a dispersion effect on the Fe_xO_y/SBA-15 system. Interactions between Fe(III) and Mo(VI) ions during synthesis resulted in higher dispersed and smaller iron oxidic species compared to those without molybdenum addition. These smaller iron oxidic species with strong interactions to the surface of SBA-15 possessed a lower reducibility. Decreased reducibility with decreased iron species size was already observed for Fe_xO_y/SBA-15 samples obtained from nitrate precursor. [86] Moreover, molybdenum addition and thereby formed Fe-O-Mo structure units may further inhibit reducibility of the Fe_xO_y system because of a charge transfer between iron and molybdenum species. A postulated electron transfer from iron to oxygen, and further to molybdenum [60] yielded a strengthened Fe-O bond, and thus, hindered the reducibility. The rising shoulder at the first TPR maxima at atomic ratios between 0.07Mo/1.0Fe and 0.21Mo/1.0Fe was assigned to the bimodal particle size distribution of $Mo_xO_y\_Fe_xO_y$/SBA-15 (Figure 7.4). Smaller Fe(III) oxidic species resulting from the dispersion effect of molybdenum were correlated with the shoulder at the first TPR maxima at higher temperatures. However, presence of Fe-O-Mo units in small oligomeric species might also influence the first TPR maxima. Because of an excess of iron compared to molybdenum, reduction of Fe(III) oxidic species to Fe(II) oxidic species was still ascribed to the first TPR maxima. Besides bimodal particle size distribution and correlated rising shoulder at the first TPR maxima, the shift of these maxima towards higher temperatures was also induced by the dispersion effect of molybdenum on the Fe_xO_y/SBA-15 system.

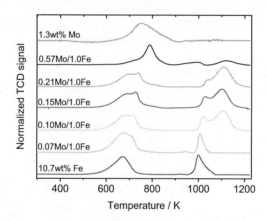

Figure 7.9: TPR traces of $Mo_xO_y\_Fe_xO_y$/SBA-15 samples at various Mo/Fe atomic ratios, 1.3wt% Mo_SBA-15, and 10.7wt% Fe_Citrate, measured in 5% H_2 in argon at a heating rate of 10 K/min.

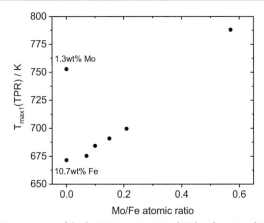

Figure 7.10: Temperature of the first TPR maxima, $T_{max1}(TPR)$, as function of Mo/Fe atomic ratio. Samples 1.3wt% Mo_SBA-15 and 10.7wt% Fe_Citrate are depicted at Mo/Fe atomic ratio of 0. TPR experiments were performed in 5% H_2 in argon at a heating rate of 10 K/min.

The second TPR maxima were also affected by molybdenum addition. Increasing Mo/Fe atomic ratio induced a decreasing intensity and an increasing shift of the second TPR maxima to higher temperatures, while a third TPR maximum arose for atomic ratios up from 0.10Mo/1.0Fe. This new TPR maximum at higher temperatures was attributed to the formation of small, hardly reducible Fe-O-Mo structure units, possibly $Fe_2(MoO_4)_3$-like species. Boulaoued et al. [186] reported for Fe-Mo/KIT-6 strong interactions between iron and molybdenum in mixed oxide catalysts, resulting in a changed reduction mechanism and a more difficult reduction of the iron oxidic species. Interestingly, they ascribed a more important role to molybdenum compared to iron. An analogous influence of molybdenum addition was reported by Kharaji et al. [59] for Fe-Mo/Al$_2$O$_3$ catalysts. The authors correlated the new third reduction peak with the presence of hardly reducible Fe-Mo composite oxides, such as $Fe_2(MoO_4)_3$. These findings corroborate the assumed strong interactions between iron and molybdenum, and the involved formation of Fe-O-Mo structure units, in the Mo_xO_y_Fe_xO_y/SBA-15 catalysts.

TPR traces of sample 1.3wt% Mo_SBA-15 differed significantly from those of Mo_xO_y_Fe_xO_y/SBA-15 catalysts and also from those reported for bulk MoO_3. [186, 187] This sample possessed only one single TPR peak. The broad TPR peak was indicative of a high dispersion of small and mainly mononuclear Mo_xO_y species, as observed by XRD and DR-UV-Vis spectroscopy. Furthermore, high reduction temperature of 1.3wt% Mo_SBA-15 was ascribed to the small Mo_xO_y species possessing strong interactions to the support material.

Long-range ordered structure after reduction

Wide-angle diffraction patterns of all $Mo_xO_y\_Fe_xO_y$/SBA-15 samples as well as samples 1.3wt% Mo_SBA-15 and 10.7wt% Fe_Citrate, after TPR measurements recorded at ambient temperature in air, are depicted in Figure 7.11. The broad diffraction peak at 23° 2θ, corresponding to amorphous SBA-15, was retained after reduction experiments for all samples. Additionally, a weak diffraction peak at 44° 2θ was detected for atomic ratios between 0.07Mo/1.0Fe and 0.21Mo/1.0Fe and for 10.7wt% Fe_Citrate. This indicated a complete reduction of Fe(III) oxidic species to Fe(0). Conversely, sample 0.57Mo/1.0Fe/SBA-15 showed a weak diffraction peak at 41° 2θ, indicative of the formation of Mo(0). The absence of diffraction peaks in the XRD pattern of 1.3wt% Mo_SBA-15 was in accordance with the low molybdenum loading and the highly dispersed Mo_xO_y species.

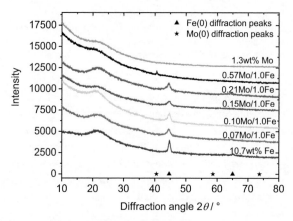

Figure 7.11: Wide-angle X-ray diffraction patterns of $Mo_xO_y\_Fe_xO_y$/SBA-15 catalysts at various Mo/Fe atomic ratios, 10.7wt% Fe_Citrate, and 1.3wt% Mo_SBA-15 recorded after TPR measurements at ambient temperature in air. Characteristic diffraction peaks of Fe(0) and Mo(0) are depicted as triangles and stars, respectively.

Formation of Mo(0) for sample 0.57Mo/1.0Fe/SBA-15 revealed the strong influence of molybdenum addition on Fe_xO_y/SBA-15 catalyst. At lower Mo/Fe atomic ratios, both dispersion and electronic effect of molybdenum addition seemed to be predominant for determining the reducibility. However, higher atomic ratio additionally led to formation of Fe-O-Mo structure units. Assuming partial formation of $Fe_2(MoO_4)_3$-like species was in accordance with the three-step reduction mechanism and the Mo(0) diffraction peak for sample 0.57Mo/1.0Fe/SBA-15.

7.4.2 Catalytic performance in selective oxidation of propene

Catalytic performance of $Mo_xO_y\_Fe_xO_y$/SBA-15 catalysts in selective oxidation of propene was investigated at 653 K and at comparable propene conversions between 5 and 8%. Figure 7.12 depicts the evolution of acrolein selectivity as function of time on stream at 653 K for $Mo_xO_y\_Fe_xO_y$/SBA-15 catalysts and 10.7wt% Fe_Citrate, respectively. Acrolein selectivity and further product distribution for sample 1.3wt% Mo_SBA-15 were omitted from the figure. For this low loaded sample, propene conversion was too low to yield reliable results. In order to ensure the absence of gas phase diffusion limitations in selective oxidation of propene, the Weisz-Prater criterion [164] was used as described in chapter 5.5.2. Due to $C_{WP} < 1$, gas phase diffusion effects were neglected. Furthermore, Koros-Nowak test [165] (see chapter 5.5.2) was applied to ensure elimination of heat and mass transport limitations in catalytic reaction.

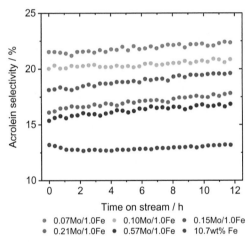

Figure 7.12: Acrolein selectivity of $Mo_xO_y\_Fe_xO_y$/SBA-15 catalysts at various Mo/Fe atomic ratios and 10.7wt% Fe_Citrate as function of time on stream at 653 K in 5% propene and 5% oxygen in helium.

Acrolein selectivity showed a significant increase due to molybdenum addition. At an atomic ratio of 0.07Mo/1.0Fe, acrolein selectivity was increased by a factor of 1.7. A further increase in Mo/Fe atomic ratio induced a decrease in acrolein selectivity. However, all $Mo_xO_y\_Fe_xO_y$/SBA-15 catalysts possessed higher acrolein selectivity compared to corresponding Fe_xO_y/SBA-15 catalyst (Figure 7.13, left, top). Besides acrolein, carbon oxides, CO and CO_2, were the main reaction products in selective oxidation of propene using $Mo_xO_y\_Fe_xO_y$/SBA-15 catalysts. Selectivity towards total oxidation products, CO_x, was significantly decreased after molybdenum addition (Figure 7.13, left, bottom). Lowest acrolein selectivity and highest CO_x selectivity for 10.7wt% Fe_Citrate was in accordance with the largest Fe_xO_y species size and the highest reducibility compared to $Mo_xO_y\_Fe_xO_y$/SBA-15 catalysts. Obviously, Fe-O bond strength without molybdenum addition was weaker leading to a favored total oxidation of propene. The lower reducibility of the Mo_xO_y_Fe_xO_y/SBA-15 catalysts, and the increased Fe-O bond strength, correlated with an increased acrolein selectivity and a lower amount of total oxidation products. Furthermore, formation of by-products, e.g. partial degradation products such as acetaldehyde and acetic acid, was observed and corresponding selectivities were determined. Formation of by-products might be explained by considering various possible reaction pathways for selective oxidation of propene. Assuming Mo_xO_y_Fe_xO_y/SBA-15 catalysts exhibit reaction pathways similar to those postulated for bulk mixed metal oxides [10, 168], selective oxidation of propene might proceed in three different pathways (I, II, and III) as shown in Figure 7.14. First, three different alcohols might be formed as intermediates. Afterwards, corresponding aldehydes, acrolein (I), propionaldehyde (II), or acetone (III) seemed likely to result from oxidation of corresponding alcohol intermediates. Consecutive oxidation may lead to acrylic acid (I), acetaldehyde (II), or acetic acid (III). Formation of acrylic acid was not observed for Mo_xO_y_Fe_xO_y/SBA-15 and 10.7wt% Fe_Citrate catalysts. Selectivities towards 1-propanol, propionaldehyde, and acetaldehyde were summarized and depicted as function of Mo/Fe atomic ratio in Figure 7.13 (right, top). These reaction products represented possible reaction pathway II. The sum of selectivities towards isopropanol, acetone, and acetic acid, representing reaction pathway III, is also depicted in Figure 7.13 (right, bottom). Molybdenum addition affected an increase in selectivities towards isopropanol, acetone, and acetic acid, and selectivities towards 1-propanol, propionaldehyde, and acetaldehyde. Increasing Mo/Fe atomic ratio correlated with further increasing selectivities towards isopropanol, acetone, and acetic acid and decreasing selectivities towards 1-propanol, propionaldehyde, and acetaldehyde. This was indicative of a change in reaction pathway dependent on Mo/Fe atomic ratio. For 10.7wt% Fe_Citrate, both sum of selectivities representing reaction pathway II and III were low. Hence, selective oxidation of propene proceeded mainly according to reaction pathway I and a high amount of total oxidation products seemed likely for this sample. Molybdenum addition yielded a significant decrease of total oxidation products, whereas reaction pathway II became more likely. Increasing Mo/Fe atomic ratio induced an increasing probability of reaction pathway III compared to reaction pathway II. Acetic acid was the main reaction product of pathway III, resulting from oxidation of isopropanol and further oxidation of acetone. According to reaction pathway II, formation of acetaldehyde resulted from oxidation of 1-propanol and further of propionaldehyde. Apparently, formation of acetaldehyde and acetic acid might be very sensitive to metal oxide species and their size. A structure sensitivity of partial oxidation reactions on metal oxides, e.g. on MoO_3, was already reported in literature. [146, 170]

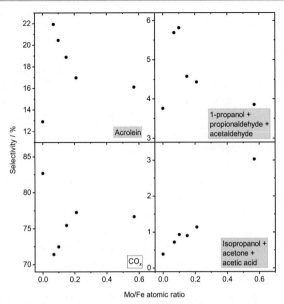

Figure 7.13: Left: Selectivity towards acrolein (top) and CO_x (bottom) as function of Mo/Fe atomic ratio. Right: Sum of selectivities towards minor reaction products according to reaction pathway II (top) and according to reaction pathway III (bottom) as function of Mo/Fe atomic ratio. Corresponding reaction pathways are illustrated in Figure 7.14.

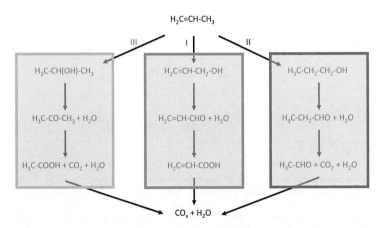

Figure 7.14: Main reaction pathways in selective oxidation of propene (adapted from [10, 168]).

7.5 Correlation between molybdenum induced effects and catalytic performance

As starting point for deducing reliable correlations between molybdenum addition and catalytic performance, acrolein selectivity, DR-UV-Vis edge energy, and temperature of first TPR maxima as function of Mo/Fe atomic ratio are depicted in Figure 7.15. Molybdenum addition induced a dispersion effect on supported iron oxidic species as explained in chapter 7.3.2. This dispersion effect, and furthermore, resulting smaller, higher dispersed Fe_xO_y species could be a possible explanation for the observed catalytic performance of the $Mo_xO_y\_Fe_xO_y$/SBA-15 catalysts. DR-UV-Vis edge energy was used as a measure for particle size of $Mo_xO_y\_Fe_xO_y$/SBA-15 catalysts. The significantly increased DR-UV-Vis edge energy due to molybdenum addition from 2.67 eV for 10.7wt% Fe_Citrate to 2.80 eV for 0.07Mo/1.0Fe_SBA-15 correlated with a significantly increased acrolein selectivity (Figure 7.15, left). However, the subsequent decrease in acrolein selectivity with increasing Mo/Fe atomic ratio could not be explained merely based on a dispersion effect. With increasing Mo/Fe atomic ratio, the influence of the electronic effect of molybdenum addition seemed to be crucial for the observed trend in acrolein selectivity. With increasing molybdenum addition, increasing Fe-O bond strength and decreasing reducibility was revealed for $Mo_xO_y\_Fe_xO_y$/SBA-15 catalysts (chapter 7.4.1). Decreasing reducibility, expressed as increasing temperature of the first TPR maxima (Figure 7.15, right), correlated with a decreasing acrolein selectivity. These observations agreed with the postulated intermediate metal-oxygen bond strength being required for selective oxidation catalysts. [7]

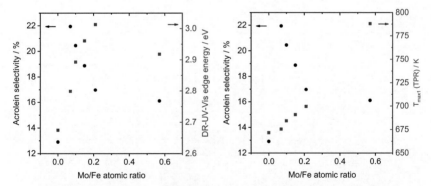

Figure 7.15: Correlation between acrolein selectivity and DR-UV-Vis edge energy (left) and temperature of first TPR maxima (right) as function of Mo/Fe atomic ratio. Acrolein selectivity was measured at 653 K in 5% propene and 5% oxygen in helium, DR-UV-Vis edge energy was determined at ambient temperature in air, and TPR measurements were conducted in 5% H_2 in argon at 10 K/min up to 1223 K.

Reaction rate of propene conversion as function of Mo/Fe atomic ratio (Figure 7.16) clarified the varying influence of dispersion effect, electronic effect, formation of Fe-O-Mo structure units, and change in reaction pathway. In general, for selective oxidation reactions spatially isolated active sites on the surface of the support material are assumed to be relevant for selectivity. [7, 27] An increasing dispersion effect with increasing Mo/Fe atomic ratio up to 0.10Mo/1.0Fe was in accordance with the postulated increased number of active sites and increased selectivity towards selective oxidation products at the expense of reaction rate. However, molybdenum addition induced a change in reaction pathway from pathway I towards pathway II (Figure 7.14) yielding an increased acetaldehyde selectivity (Figure 7.13). Exceeding the atomic ratio of 0.10Mo/1.0Fe resulted in an increased reaction rate, while acrolein selectivity decreased. Probability for reaction pathway II significantly decreased for an atomic ratio of 0.15Mo/1.0Fe compared to 0.10Mo/1.0Fe. The further increased dispersion effect at 0.21Mo/1.0Fe resulted in both decreased reaction rate and decreased acrolein selectivity, whereas selectivity towards acetic acid increased. At this atomic ratio, the higher amount of Fe-O-Mo structure units as well as the effect of decreased reducibility seemed to become more influential than the effect of dispersion. At 0.57Mo/1.0Fe, the dispersion effect still existed but the electronic effect was predominant. Probability for reaction pathway III significantly increased at the expense of acrolein selectivity. Furthermore, formation of hardly reducible Fe-O-Mo structure units resulted in a significantly decreased reaction rate.

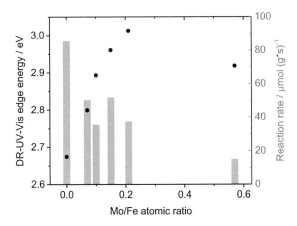

Figure 7.16: DR-UV-Vis edge energy (left axis, dots) and reaction rate of propene conversion (right axis, bars) as function of Mo/Fe atomic ratio. Reaction rates were measured in 5% propene and 5% oxygen in helium at 653 K.

This combination of various effects induced by molybdenum addition and catalytic performance is shown in Figure 7.17. Figure 7.17 illustrates the correlation between reducibility, particularly electronic effect of molybdenum addition, dispersion effect, expressed as DR-UV-Vis edge energy, reaction rate of propene conversion, and acrolein selectivity.

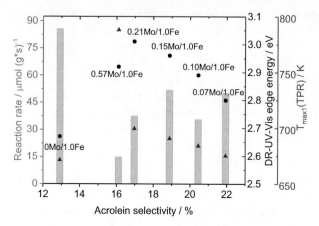

Figure 7.17: Reaction rate of propene conversion (left axis, grey bars), DR-UV-Vis edge energy (right axis, black dots) and temperature of first TPR maxima (outer right axis, blue triangles), respectively, as function of acrolein selectivity. Catalytic measurements were conducted in 5% propene and 5% O_2 in helium at 653 K, while temperature-programmed reduction experiments were performed in 5% H_2 in argon up to 1223 K.

DR-UV-Vis edge energy as function of acrolein selectivity showed a volcano-type behavior. Smallest species size at an atomic ratio of 0.21Mo/1.0Fe correlated with an intermediate acrolein selectivity. Decreasing Mo/Fe atomic ratio and decreasing DR-UV-Vis edge energy coincided with an increasing acrolein selectivity and a decreasing temperature of first TPR maxima. Conversely, increasing atomic ratio up to 0.57Mo/1.0Fe induced a decreased DR-UV-Vis edge energy and a distinctly increased temperature of first TPR maxima, while acrolein selectivity decreased.

In contrast to DR-UV-Vis edge energy and temperature of first TPR maxima, reaction rate of propene conversion followed no clear trend. Up to 16.1% acrolein selectivity reaction rate significantly decreased. Beyond this point an increasing reaction rate up to 18.9% acrolein selectivity was observed. Exceeding an acrolein selectivity of 18.9% induced a decrease in reaction rate, whereas further increasing acrolein selectivity up to 21.9% correlated with a significantly increased reaction rate. Hence, dispersion and size of supported iron oxidic species on SBA-15 as well as reducibility influenced the catalytic performance in selective oxidation of propene. Not only one of the molybdenum induced effects on Fe_xO_y/SBA-15 catalysts seemed to be responsible for the catalytic performance, but rather a combination of all effects.

7.6 Interpretation of molybdenum induced differences in structure and reducibility of Fe$_x$O$_y$/SBA-15 catalysts and their impact on catalytic performance

Figure 7.18 depicts a schematic representation of the main structural motifs of Mo$_x$O$_y$_Fe$_x$O$_y$/SBA-15 catalysts dependent on Mo/Fe atomic ratio. Based on this schematic representation, the influence of molybdenum addition on Fe$_x$O$_y$/SBA-15 catalysts, i.e. on their structure, reducibility, and catalytic performance, might be interpreted.

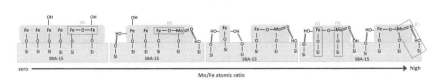

Figure 7.18: Schematic representation of structural differences of iron oxidic species supported on SBA-15 dependent on Mo/Fe atomic ratio. Only the main structural motifs of Mo$_x$O$_y$_Fe$_x$O$_y$/SBA-15 catalysts are depicted.

Increasing Mo/Fe atomic ratio correlated with a decreasing amount of Fe-O-Fe structure units (Figure 7.18, (A)). This was ascribed to smaller, higher dispersed Fe$_x$O$_y$ species of Mo$_x$O$_y$_Fe$_x$O$_y$/SBA-15 catalysts compared to those of corresponding Fe$_x$O$_y$/SBA-15 catalyst without molybdenum. The dispersion effect of molybdenum on Fe$_x$O$_y$ species induced this decreasing species size. Since bridging oxygen atoms in Fe-O-Fe structure units were presumably required for reducibility of Fe$_x$O$_y$/SBA-15 catalysts, the decreased Fe$_x$O$_y$ species size correlated with a decreased reducibility of Mo$_x$O$_y$_Fe$_x$O$_y$/SBA-15 catalysts compared to corresponding Fe$_x$O$_y$/SBA-15 catalyst. Furthermore, with increasing Mo/Fe atomic ratio, Fe-O-Fe structure units were partially replaced by Fe-O-Mo structure units (Figure 7.18, (B)). Accordingly, the electronic effect of molybdenum on Fe$_x$O$_y$ species became more important for their reducibility. This electronic effect was associated with an electron transfer from iron to oxygen, and further to molybdenum in Fe-O-Mo structure units. As a result, Fe-O bonds between iron and bridging oxygen atoms were strengthened, yielding a lower reducibility (Figure 7.18, (B)). Only for highest Mo/Fe atomic ratio, an influence of molybdenum addition on micropore filling was observed. This high Mo/Fe atomic ratio induced a higher degree of micropore filling being ascribed to a higher amount of Fe-O-Si (Figure 7.18, (C)) and Mo-O-Si bonds (Figure 7.18, (D)), as well as hydrogen bonds (Figure 7.18, (E)). These hardly reducible Fe-O-Si and Mo-O-Si bonds additionally hindered the reducibility of Mo$_x$O$_y$_Fe$_x$O$_y$/SBA-15 catalysts. Therefore, selectivity towards acrolein decreased with increasing Mo/Fe atomic ratio. Highest acrolein selectivity for an atomic ratio of 0.07Mo/1.0Fe might be explained by an optimal combination of intermediate reducibility and high electron transfer efficiency in Fe$_x$O$_y$ oligomers. This might facilitate the formation of nucleophilic surface oxygen species, yielding a higher acrolein selectivity. Despite decreasing reducibility with increasing Mo/Fe atomic ratio, formation of nucleophilic oxygen species might still be favored in Mo$_x$O$_y$_Fe$_x$O$_y$/SBA-15 catalysts. This might explain the higher acrolein selectivity of all

$Mo_xO_y\_Fe_xO_y$/SBA-15 catalysts compared to that of corresponding Fe_xO_y/SBA-15 catalyst. However, acrolein selectivity as function of Mo/Fe atomic ratio was presumably determined by a combination of various effects, such as reducibility, species size, electron transfer efficiency, and also change in reaction pathway. Characterization of $Mo_xO_y\_Fe_xO_y$/SBA-15 catalysts corroborated the assumed importance of Fe-O-Fe structure units (Figure 7.18, (A)) in selective oxidation of propene.

7.7 Conclusion

Synthesis of iron and molybdenum mixed oxides on mesoporous SBA-15 yielded suitable binary model catalysts for investigating structure-activity correlations. $Mo_xO_y\_Fe_xO_y$/SBA-15 catalysts were characterized by a multitude of analyzing methods. Supporting iron and molybdenum oxides on SBA-15 by incipient wetness technique led to the desired atomic ratios between 0.07Mo/1.0Fe and 0.57Mo/1.0Fe. Invariant iron loading permitted investigations of the influence of molybdenum addition on Fe_xO_y/SBA-15 catalysts. Furthermore, for comparison to Mo_xO_y/SBA-15 system, sample 1.3wt% Mo_SBA-15, consisting of mainly isolated mononuclear Mo_xO_y species, was successfully synthesized. Mo_xO_y and Fe_xO_y species of the $Mo_xO_y\_Fe_xO_y$/SBA-15 catalysts were highly dispersed on SBA-15 without changing the structure of the support. Formation of long-range ordered structures was excluded. Interactions between iron and molybdenum induced a bimodal particle size distribution at atomic ratios between 0.07Mo/1.0Fe and 0.21Mo/1.0Fe due to an excess of iron compared to molybdenum. Higher Mo/Fe atomic ratio yielded a broadened particle size distribution. Furthermore, low molybdenum addition, at atomic ratios up to 0.21Mo/1.0Fe, did neither affect the contribution of micropores to the entire surface area nor the surface roughness of SBA-15. Conversely, an atomic ratio of 0.57Mo/1.0Fe resulted in a decreased contribution of micropores to the entire surface area of SBA-15. This indicated an enhanced micropore filling, complying with a smoother support surface. Low molybdenum loading of sample 1.3wt% Mo_SBA-15, and therefore, highly dispersed Mo_xO_y species, was in accordance with a higher contribution of micropores to the entire surface area of SBA-15 and a rougher SBA-15 surface. Molybdenum addition yielded a pronounced dispersion effect on supported iron oxidic species. Increasing atomic ratio up to 0.21Mo/1.0Fe was accompanied by decreasing species sizes. Strong interactions between iron and molybdenum during synthesis resulted in formation of Fe-O-Mo structure units. At higher Mo/Fe atomic ratio, these strong interactions probably induced small $Fe_2(MoO_4)_3$-like species.

Reducibility of $Mo_xO_y\_Fe_xO_y$/SBA-15 catalysts was investigated by temperature-programmed reduction experiments with hydrogen as reducing agent. Molybdenum addition significantly influenced the reducibility of Fe_xO_y/SBA-15. Lower reducibility due to molybdenum addition was ascribed to both dispersion and electronic effect of molybdenum. Smaller and higher dispersed iron oxidic species possessed a lower reducibility compared to larger and less dispersed species. Additionally, a charge transfer from iron to oxygen, and further to molybdenum in Fe-O-Mo structure units yielded a strengthened Fe-O bond, and hence, hindered reducibility. The change of the two-step reduction mechanism for 10.7wt% Fe_Citrate towards a three-step reduction mechanism for $Mo_xO_y\_Fe_xO_y$/SBA-15 catalysts corroborated the presence of small, hardly reducible Fe-O-Mo structure units, possibly $Fe_2(MoO_4)_3$-like species.

Catalytic performance of $Mo_xO_y\_Fe_xO_y$/SBA-15 was studied under selective propene oxidation conditions. Molybdenum addition resulted in an increased acrolein selectivity and a decreased selectivity towards total oxidation products. Increasing Mo/Fe atomic ratio induced a decreasing acrolein selectivity being however higher than that of Fe_xO_y/SBA-15 catalyst without molybdenum addition. Besides, increasing Mo/Fe atomic ratio resulted in an increased selectivity towards acetaldehyde and acetic acid. This was indicative of a change in reaction pathway. Influence of molybdenum addition on catalytic performance was correlated with both dispersion and electronic effect of molybdenum. The strengthened Fe-O bonds, and the lower reducibility, with increasing Mo/Fe atomic ratio, led to an inferior total oxidation. This coincided with an intermediate Fe-O bond strength as postulated being required for selective oxidation of propene. Correlation of reaction rate of propene conversion, acrolein selectivity, DR-UV-Vis edge energy, and temperature of first TPR maxima clarified the varying influence of electronic effect, dispersion effect, and formation of Fe-O-Mo structure units.

8 Summary and outlook

8.1 Introduction

Understanding the correlation between structure and catalytic performance of a catalyst is an important issue in catalysis research. Revealing reliable structure-activity correlations requires reducing chemical and structural complexity of industrial applied catalysts to simplified model catalysts. In heterogeneous catalysis, catalytic reactions occur on the surface of the catalysts, while the surface structure differs significantly from that of the bulk. Therefore, dispersing metal oxides on well-defined support materials may result in suitable model catalysts. For elucidating structure-activity correlations, a detailed knowledge of structure, composition, and certain chemical functions of the model catalysts is indispensable. Thus, determining the important variables for structure and catalytic performance of the model catalysts is a starting point for a rational design of improved catalysts. The objective of this work was synthesizing and characterizing iron oxidic species supported on SBA-15 as model catalysts for selective oxidation of propene. In addition to *ex situ* and *in situ* characterization, applicability of solid-state kinetic analysis was shown for supported iron oxidic species. Results from solid-state kinetic analysis under non-isothermal conditions were particular helpful in corroborating structure-activity correlations of Fe_xO_y/SBA-15 catalysts. Investigations of Fe_xO_y/SBA-15 catalysts focused on the influence of iron loading, Fe(III) precursor, and powder layer thickness during calcination on resulting iron oxidic species, their reducibility, and catalytic performance. Subsequently, a binary model catalyst system was synthesized by changing the chemical composition of the Fe_xO_y/SBA-15 catalysts by molybdenum addition. Therefore, iron and molybdenum mixed oxides supported on SBA-15 were synthesized and characterized regarding structure, reducibility, and catalytic performance in selective oxidation of propene.

8.2 Support material

In heterogeneous catalysis, suitable support materials have to possess high specific surface area, uniform pore structure, high stability, and inactivity in investigated reaction. In this work, SBA-15 was investigated regarding its applicability as support material for iron oxidic species as model catalysts for selective oxidation of propene. *Ex situ* characterization confirmed SBA-15 fulfilling structural requirements for support materials. Moreover, all synthesized SBA-15 samples possessed not only comparable specific surface areas, but also comparable values of lattice constant and wall thickness. Accordingly, synthesis of SBA-15 was successful and reproducible. *In situ* investigations of SBA-15 clarified high stability and inactivity in selective oxidation of propene. Consequently, SBA-15 was proven to be a suitable support material for iron oxide-based catalysts in this work.

8.3 Structure of Fe_xO_y/SBA-15 catalysts

Iron oxidic species supported on SBA-15 were synthesized by incipient wetness technique using two different Fe(III) precursors, $(NH_4, Fe(III))$ citrate and Fe(III) nitrate nonahydrate. Iron loading ranged from 2.0 through 10.7wt% yielding surface coverages of 0.30 to 1.90 Fe atoms per nm^2. Synthesis procedure led to highly dispersed and small iron oxidic species in the pore system of SBA-15. Characteristic two-dimensional hexagonal pore structure of SBA-15 was retained during synthesis. Formation of long-range ordered phases and monolayer coverage was excluded for all Fe_xO_y/SBA-15 catalysts.

Increasing iron loading induced an increasing iron oxidic species size. Iron oxidic species on SBA-15 were predominantly dimeric or small Fe_xO_y oligomers with octahedrally coordinated Fe^{3+}. For all Fe_xO_y/SBA-15 catalysts, presence of Fe_xO_y species with a local geometry similar to that of α-Fe_2O_3 or β-FeOOH was revealed. Varying iron loading was not only crucial for resulting iron oxidic species size, but also for amount of β-FeOOH-like species on SBA-15. Decreasing iron loading correlated with increasing amount of β-FeOOH-like species.

Fe(III) precursor significantly influenced species size, surface and porosity characteristics, and types of iron oxidic species of Fe_xO_y/SBA-15 catalysts. Precursor-dependent structural differences of Fe_xO_y/SBA-15 catalysts were ascribed to the different strengths of chelating effect of the two precursors. The citrate precursor, possessing a more pronounced chelating effect compared to the nitrate precursor, induced smaller, higher dispersed, and less aggregated iron oxidic species on SBA-15. These smaller iron oxidic species were accompanied by a higher degree of micropore filling and a smoother SBA-15 surface. This was indicative of stronger interactions between iron oxidic species and support material compared to corresponding nitrate samples.

8.4 Chemical memory effect of Fe_xO_y/SBA-15 catalysts

Resulting types of iron oxidic species on SBA-15 were dependent on the precursor. Fe_xO_y species possessed a local geometry similar to that of α-Fe_2O_3 or β-FeOOH. The citrate samples consisted of significantly higher amounts of β-FeOOH-like species compared to corresponding nitrate samples. These precursor-dependent differences in types of iron oxidic species were explained by a chemical memory effect of Fe_xO_y/SBA-15 catalysts. Concerning a chemical memory effect, the precursor chemistry during synthesis controls morphology and structure of the final system. Accordingly, the different strengths of chelating effect of the two precursors contributed to the different chemistry during synthesis, and further, to the different types of iron oxidic species on SBA-15. Due to the more pronounced chelating effect of the citrate precursor, precursor molecules were highly dispersed within the pore system of SBA-15 during synthesis of Fe_xO_y/SBA-15 catalysts. Hence, hydrogen bonds between citrate precursor molecules and superficial silanol groups of SBA-15 were formed. This yielded a significant amount of β-FeOOH-like species on SBA-15 after calcination. Conversely, the nitrate precursor with a less pronounced chelating effect, induced minor amounts of hydrogen bonds between precursor molecules and superficial silanol groups of SBA-15. These differences in amount of hydrogen bonds between precursor molecules and SBA-15 during synthesis presumably induced the different amounts of β-FeOOH-like species on SBA-15.

8.5   Applicability of solid-state kinetic analysis

For all Fe_xO_y/SBA-15 catalysts, solid-state kinetic analysis of non-isothermal reduction was successfully applied. Moreover, results from solid-state kinetic analysis corroborated those from structural characterization of Fe_xO_y/SBA-15 catalysts. Both iron loading and precursor influenced solid-state kinetic properties of supported iron oxidic species. For iron loadings higher than 2.5wt%, citrate samples possessed higher values of apparent activation energy of rate-determining step in reduction compared to corresponding nitrate samples. This was ascribed to smaller iron oxidic species obtained from the citrate precursor, accompanied by a higher degree of micropore filling and a smoother SBA-15 surface. These surface and porosity characteristics were indicative of stronger interactions between iron oxidic species and SBA-15. Thus, reducibility was hindered. Within the citrate samples, decreasing iron loading induced a decreasing apparent activation energy in reduction. This coincided with higher amounts of better reducible β-FeOOH-like species and a lower degree of micropore filling. Conversely, within the nitrate samples, decreasing iron loading correlated with an increasing apparent activation energy in reduction. Lower apparent activation energy in reduction for higher loaded nitrate samples was ascribed to weaker interactions between iron oxidic species and SBA-15, and moreover, coincided with iron loading-dependent particle size effect. Model-dependent Coats-Redfern method identified the first-order, Mampel, model as suitable solid-state reaction model for describing the reduction of Fe_xO_y/SBA-15 catalysts. The Mampel model describes solid-state reactions with a large number of nucleation sites resulting in fast nucleation. This agreed well with the small iron oxidic species being highly dispersed in the pore system of SBA-15 and weakly interacting with each other. Additionally, JMAK kinetic analysis was consistent with one-dimensional reduction of small iron oxidic species located in the pore system of SBA-15. Only for the highest loaded nitrate sample, the dimensionality of reduction increased. This agreed well with the largest iron oxidic species and the highest amount of α-Fe_2O_3-like species for this sample.

8.6   Structure-activity correlations of Fe_xO_y/SBA-15 catalysts dependent on precursor

Catalytic performance of Fe_xO_y/SBA-15 catalysts was investigated during selective oxidation of propene to acrolein. Used precursor crucially determined the catalytic performance of Fe_xO_y/SBA-15 catalysts. The citrate precursor induced smaller Fe_xO_y species and higher amounts of β-FeOOH-like species compared to the nitrate precursor. This resulted in both a higher reaction rate of propene conversion and a higher acrolein selectivity for the citrate samples up to an iron loading of 7.2wt%. Lowest loaded citrate and nitrate sample possessed the highest reaction rate of propene conversion, being ascribed to the highest amount of β-FeOOH-like species rather than the smallest Fe_xO_y species size. Fe_xO_y/SBA-15 catalysts obtained from citrate precursor and those obtained from nitrate precursor showed an opposite trend in acrolein selectivity as function of iron loading. For citrate samples, increasing acrolein selectivity correlated with decreasing iron loading. Conversely, for nitrate samples, increasing acrolein selectivity correlated with increasing iron loading. This opposite trend in acrolein selectivity was ascribed to a precursor-dependent difference in influence of reducibility and phase composition. Catalytic performance of Fe_xO_y/SBA-15 catalysts obtained from citrate precursor directly correlated with the amount of β-FeOOH-like species. For these catalysts, onset of acrolein formation started at an amount of 50% of β-FeOOH-like species under propene oxidation conditions. Within the citrate samples, decreasing iron loading correlated with increasing amount of β-FeOOH-like species, and hence, with an increasing selectivity towards acrolein. Highest acrolein selectivity for lowest loaded citrate samples might be induced by a favored H abstraction of the methyl group of propene due to the high amount of nucleophilic hydroxyl groups of Fe-OH structure units. Furthermore, the high amount of β-FeOOH-like species, and therefore, of neighboring hydroxyl groups might facilitate the recombination of hydroxyl groups and abstracted H from propene to form water and a reduced iron center. Conversely, Fe_xO_y/SBA-15 catalysts obtained from nitrate precursor consisted mainly of α-Fe_2O_3-like species. For these catalysts, influence of surface and porosity characteristics and reducibility was predominantly determining the catalytic performance. Acrolein selectivity increased with increasing iron loading. This correlated with increasing iron oxidic species size, increasing contribution of micropores to entire surface area, increasing SBA-15 surface roughness, and increasing reducibility. Hence, weaker interactions between iron oxidic species and SBA-15, i.e. lower amounts of Fe-O-Si bonds, and higher amounts of Fe-O-Fe structure units resulted in a lower apparent activation energy in reduction. Accordingly, formation of nucleophilic O^{2-} species might be facilitated which lead to nucleophilic attacking propene at the methyl group. Moreover, better reducibility of the nitrate samples coincided with higher amounts of total oxidation products compared to corresponding citrate samples. This was in accordance with the postulated Mars-van-Krevelen or redox-type mechanism for selective oxidation of propene.

8.7 Influence of powder layer thickness during calcination on structural and solid-state kinetic properties of Fe_xO_y/SBA-15 catalysts

Calcination in either thin (0.3 cm) or thick (1.3 cm) powder layer significantly influenced structural and solid-state kinetic properties of Fe_xO_y/SBA-15 catalysts. Additionally, the magnitude of the influence of powder layer thickness during calcination was dependent on both precursor and iron loading. Detailed structural characterization of Fe_xO_y/SBA-15_Th catalysts confirmed iron oxidic species being successfully located in the pore system of SBA-15. Characteristic pore structure of SBA-15 was retained during synthesis. Thick layer calcination induced an extended retention time of gaseous decomposition products of the precursor. This yielded higher concentrations of gaseous decomposition products in the powder layer. Consequently, the equilibrium during thermal decomposition of the precursor was shifted to the side of the educts, according to Le Chatelier's principle. Hence, iron oxidic species growth was enhanced, and resulting iron oxidic species were larger compared to those obtained from thin layer calcination. The nitrate precursor induced a more pronounced aggregation of Fe_xO_y species during thick layer calcination compared to the citrate precursor. This was ascribed to a minor dispersion of Fe(III) ions of the nitrate precursor before calcination, which was induced by the less pronounced chelating effect. Accordingly, nitrate_Th samples consisted of larger iron oxidic species compared to citrate_Th samples with similar iron loading. Only for nitrate_Th samples with iron loadings higher than 2.0wt%, significant amounts of higher aggregated Fe_xO_y and small α-Fe_2O_3 particles were observed. Fe_xO_y/SBA-15_Th catalysts showed similar precursor-dependent and iron loading-dependent particle size effects as Fe_xO_y/SBA-15 catalysts. Moreover, precursor-dependent influence on surface and porosity characteristics was independent of layer thickness during calcination. Thick layer calcination induced both an increased degree of micropore filling and a smoother SBA-15 surface except for lowest loaded samples, 2.0wt% Fe_Nitrate_Th and 2.5wt% Fe_Citrate_Th. Influence of layer thickness during calcination on these lowest loaded samples was more pronounced compared to higher loaded samples. Iron oxidic species size of 2.0wt% Fe_Nitrate_Th and 2.5wt% Fe_Citrate_Th was significantly increased due to thick layer calcination. Furthermore, the broadened reduction peak in the TPR profiles of these lowest loaded samples complied with the increased apparent activation energy in reduction.

Reduction mechanism of citrate_Th samples remained unaffected by layer thickness during calcination. Conversely, reduction mechanism of nitrate_Th samples became more complex with less resolved reduction peaks in the multi-step TPR profiles. For iron loadings higher than 2.5wt%, citrate_Th samples possessed higher values of apparent activation energy in reduction compared to nitrate_Th samples with similar iron loading. This correlated with the precursor-dependent differences in strengths of interactions between iron oxidic species and SBA-15. Furthermore, for the citrate samples, thick layer calcination resulted neither in a change in suitable solid-state reaction model, nor in a change in geometrical characteristics of the reduction. Conversely, for the nitrate samples, significant differences in solid-state kinetic properties due to thick layer calcination were observed. For nitrate samples with iron loadings higher than 2.0wt%, thick layer calcination induced significant amounts of higher aggregated iron oxidic species and small α-Fe_2O_3 particles. This coincided with an increased reaction order of suitable solid-state reaction model and a higher dimensionality of reduction. For lowest loaded nitrate_Th sample, suitable solid-state reaction model and geometrical characteristics of the reduction were unaffected by layer thickness during calcination.

8.8 Influence of powder layer thickness during calcination on structure-activity correlations of Fe_xO_y/SBA-15 catalysts

In situ characterization of Fe_xO_y/SBA-15_Th catalysts under propene oxidation conditions revealed a significant influence of powder layer thickness during calcination on their catalytic performance. This influence was furthermore dependent on precursor and iron loading. Particle size effects determined the reaction rate of propene conversion. Higher reaction rates of propene conversion of citrate_Th compared to nitrate_Th samples were ascribed to the precursor-dependent particle size effect. Iron loading-dependent particle size effect induced a decreasing reaction rate of propene conversion with increasing iron loading. For citrate_Th samples, iron oxidic species size, strengths of interactions between iron oxidic species and SBA-15, and presumably amount of β-FeOOH-like species determined the catalytic performance in selective oxidation of propene. Interactions between iron oxidic species and SBA-15 were associated with oxygen availability. Too strong interactions between iron oxidic species and SBA-15 consequently hindered the selective oxidation of propene due to an insufficient oxygen availability. Conversely, too weak interactions yielded a too high oxygen availability, and hence, a favored total oxidation of propene. Within citrate_Th samples, acrolein selectivity increased with decreasing iron loading. This coincided with weaker interactions between iron oxidic species and SBA-15. Moreover, increased acrolein selectivity of 2.5wt% Fe_Citrate_Th compared to 2.5wt% Fe_Citrate correlated with both weaker interactions between iron oxidic species and SBA-15 and lower amounts of isolated Fe^{3+} species. Conversely, decreased acrolein selectivity of higher loaded citrate_Th samples compared to corresponding citrate samples correlated with too strong interactions between iron oxidic species and SBA-15.

Nitrate_Th samples showed a volcano-type behavior of acrolein selectivity as function of iron loading. Increasing iron loading up to 7.2wt% correlated with an increasing acrolein selectivity of nitrate_Th samples. This was ascribed to larger iron oxidic species and a lower apparent activation energy in reduction. Further increasing iron loading up to 9.3wt% coincided with a decreasing acrolein selectivity due to a significant amount of small α-Fe_2O_3 particles. All nitrate_Th samples possessed an increased acrolein selectivity compared to corresponding nitrate samples. Increased acrolein selectivity of 2.0wt% Fe_Nitrate_Th compared to 2.0wt% Fe_Nitrate correlated with larger iron oxidic species, and hence, minor amounts of isolated Fe^{3+} species. Conversely, increased acrolein selectivity of 9.3wt% Fe_Nitrate_Th compared to 9.3wt% Fe_Nitrate was predominantly ascribed to stronger interactions between iron oxidic species and support material. These stronger interactions hindered the total oxidation reaction and favored the selective oxidation pathway, according to an intermediate metal-oxygen bond strength being required for selective oxidation reactions. Increased acrolein selectivity of 7.2wt% Fe_Nitrate_Th compared to 7.2wt% Fe_Nitrate correlated with a decreased apparent activation energy in reduction and an increased iron oxidic species size. For Fe_xO_y/SBA-15_Th catalysts obtained from nitrate precursor, thick layer calcination affected not only species size and surface and porosity characteristics, but also reducibility being reflected in their catalytic performance.

8.9 Influence of molybdenum addition on mesoporous structure of Fe$_x$O$_y$/SBA-15 catalysts

Mixed iron and molybdenum oxide catalysts supported on SBA-15 were prepared by incipient wetness technique. The pH value of the aqueous oxide precursor solution was adjusted to 7.5-8. Thus, Fe(III) ions in the dissolved (NH$_4$, Fe(III)) citrate precursor directly interacted with Mo(VI) ions in the dissolved ammonium heptamolybdate tetrahydrate precursor, being present as isolated tetrahedral MoO$_4^{2-}$ species. Hence, formation of iron and molybdenum mixed oxides was preferred. To investigate the influence of molybdenum addition on Fe$_x$O$_y$/SBA-15 catalysts, iron loading was kept invariant, while Mo/Fe atomic ratio was varied between 0.07/1.0 and 0.57/1.0. Iron and molybdenum oxidic species were small and highly dispersed in the pore system of SBA-15. Characteristic pore structure of SBA-15 was retained during synthesis. Mo$_x$O$_y$_Fe$_x$O$_y$/SBA-15 catalysts showed no long-range ordered phases. However, due to the strong interactions between iron and molybdenum during synthesis, possible formation of amorphous and highly dispersed Fe$_2$(MoO$_4$)$_3$-like species seemed reasonable. For atomic ratios up to 0.21Mo/1.0Fe, bimodal pore radius distribution was observed. This was ascribed to a bimodal particle size distribution, induced by differences in chemical environment during synthesis due to an excess of iron compared to molybdenum. At these atomic ratios, molybdenum addition affected neither the contribution of micropores to the entire surface area of SBA-15 nor the surface roughness. Conversely, an increased atomic ratio of 0.57Mo/1.0Fe resulted in a broadened particle size distribution, a decreased contribution of micropores to entire surface area, and a smoother surface area.

8.10 Dispersion and electronic effect of molybdenum addition on Fe$_x$O$_y$/SBA-15 catalysts

Molybdenum addition induced a dispersion effect on supported iron oxidic species. Accordingly, iron oxide dispersion was increased, while average species size was decreased compared to Fe$_x$O$_y$/SBA-15 catalyst. Increasing atomic ratio up to 0.21Mo/1.0Fe correlated with an enhanced dispersion effect. Further increased atomic ratio of 0.57Mo/1.0Fe was accompanied by strong interactions between iron and molybdenum in resulting Fe-O-Mo structure units and possibly in small Fe$_2$(MoO$_4$)$_3$-like species. This induced a slightly increased average species size, and hence, a slightly decreased dispersion effect. However, Fe$_x$O$_y$ species remained still smaller and higher dispersed on the support material compared to those of corresponding Fe$_x$O$_y$/SBA-15 catalyst. Molybdenum induced dispersion effect was also reflected in reducibility of Mo$_x$O$_y$_Fe$_x$O$_y$/SBA-15 catalysts. Lower reducibility of Mo$_x$O$_y$_Fe$_x$O$_y$/SBA-15 catalysts, compared to corresponding Fe$_x$O$_y$/SBA-15 catalyst, was ascribed to both dispersion and electronic effect of molybdenum on Fe$_x$O$_y$ species. Hence, smaller and higher dispersed iron oxidic species with stronger interactions to SBA-15 showed a lower reducibility. Furthermore, the electronic effect of molybdenum addition could be explained by a charge transfer from iron to oxygen, and further to molybdenum in Fe-O-Mo structure units. Consequently, Fe-O bonds in Fe-O-Mo structure units were strengthened, yielding a lower reducibility. A change of the two-step reduction mechanism for corresponding Fe_xO_y/SBA-15 catalyst towards a three-step reduction mechanism for $Mo_xO_y\_Fe_xO_y$/SBA-15 catalysts at atomic ratios higher than 0.07Mo/1.0Fe was observed. This corroborated the presence of small, hardly reducible Fe-O-Mo structure units, possibly $Fe_2(MoO_4)_3$-like species.

8.11 Enhanced catalytic performance of Fe_xO_y/SBA-15 catalysts due to molybdenum addition

$Mo_xO_y\_Fe_xO_y$/SBA-15 catalysts were suitable binary model catalysts for elucidating structure-activity correlations for selective oxidation of propene. Molybdenum addition resulted in an increased acrolein selectivity at the expense of total oxidation products. Increasing Mo/Fe atomic ratio induced a decreasing acrolein selectivity being however higher than that of corresponding Fe_xO_y/SBA-15 catalyst. Enhanced catalytic performance of Fe_xO_y/SBA-15 catalysts correlated with both dispersion and electronic effect of molybdenum. The dispersion effect of molybdenum was considered responsible for the increased acrolein selectivity at an atomic ratio of 0.07Mo/1.0Fe. Accordingly, molybdenum addition led to a favored selective oxidation of propene and an inferior total oxidation. However, the decrease in acrolein selectivity for higher Mo/Fe atomic ratios could not be explained merely based on a dispersion effect. With increasing Mo/Fe atomic ratio, the influence of the electronic effect of molybdenum seemed to become more important for acrolein selectivity. The electronic effect of molybdenum strengthened the Fe-O bonds in Fe-O-Mo structure units. Increasing Fe-O bond strength and decreasing reducibility of the $Mo_xO_y\_Fe_xO_y$/SBA-15 catalysts with increasing Mo/Fe atomic ratio correlated with a decreasing acrolein selectivity. These results agreed with the postulated intermediate metal-oxygen bond strength being required for selective oxidation catalysts. Additionally, increasing Mo/Fe atomic ratio induced an increasing selectivity towards acetaldehyde and acetic acid, indicative of a change in reaction pathway in selective oxidation of propene. Catalytic performance of $Mo_xO_y\_Fe_xO_y$/SBA-15 catalysts was determined by a combination of dispersion effect, electronic effect, formation of Fe-O-Mo structure units, and change in reaction pathway.

8.12 Outlook

Rational catalyst improvement remains still challenging, not least because multiple variables, especially during synthesis procedure, influence the catalytic performance. Nevertheless, this work showed, that iron oxidic species supported on SBA-15 constitute suitable model catalysts for selective oxidation of propene. Despite the current discrepancy between industrial achieved selectivity towards acrolein and that of Fe_xO_y/SBA-15 model catalysts, this work laid the foundation of the possibility of converting the selective oxidation of propene into a *greener* process based on environmentally friendly, less expensive iron-based catalysts. Accordingly, continuative research on *greener* model catalysts is required.

This work emphasized the influence of surface and porosity characteristics, and hence, strengths of interactions between metal oxides and support material on catalytic performance. Therefore, targeted variations of surface and porosity characteristics might be a future starting point for gaining further insight into metal oxide-support interactions and their influence on catalytic performance. This could be realized for example by tailoring the pore radii of the support material or by changing the surface acidity of the support material.

Especially for small and highly dispersed iron oxidic species supported on SBA-15 obtained from (NH$_4$, Fe(III)) citrate precursor, the amount of β-FeOOH-like species directly affected the catalytic performance in selective oxidation of propene. As short-term objective, targeted synthesis of small and highly dispersed iron oxidic species, with high amounts of β-FeOOH-like species, obtained from other Fe(III) precursors with a pronounced chelating effect, could be performed. Detailed characterization of resulting iron oxidic species under propene oxidation conditions might constitute an important step towards revealing catalytically active species.

As long-term objective, chemical complexity of the model catalysts can be further increased by stepwise addition of other environmentally friendly metals besides iron and molybdenum, for instance, copper, manganese, or zinc. Subsequently, understanding structure-activity correlations of these more complex, *green* model catalysts can be achieved. Hence, future investigations of model catalysts with increased chemical complexity, including advanced analyzing methods, and therefrom deduced structure-activity correlations, still constitute a broad field in catalysis research aiming at a rational catalyst improvement.

Solid-state kinetic analysis under non-isothermal reduction conditions was successfully applied to Fe$_x$O$_y$/SBA-15 catalysts, and hence, shown to enlarge the understanding of structure-activity correlations of supported metal oxides. The next step in gaining further insight into structure-activity correlations of supported metal oxides might be application of solid-state kinetic analysis under catalytic conditions. Therefrom deduced solid-state kinetic properties of supported metal oxides might be directly correlated with their catalytic performance.

9 References

[1] *Umsätze der wichtigsten Industriebranchen in Deutschland bis 2016 | Statistik*. https://de.statista.com/statistik/daten/studie/241480/umfrage/umsaetze-der-wichtigsten-industriebranchen-in-deutschland/. Accessed 08/08/2018.

[2] *Chemieindustrie | Statista*. https://de.statista.com/statistik/studie/id/7652/dokument/chemieindustrie-statista-dossier/. Accessed 08/08/2018.

[3] GTAI. *annual business volume chemical industry germany*. http://www.gtai.de/GTAI/Content/EN/Invest/_SharedDocs/Downloads/GTAI/Industry-overviews/industry-overview-chemical-industry-in-germany-en.pdf?v=2. Accessed 08/08/2018.

[4] Saygin, D., Worrell, E., Tam, C., Trudeau, N., Gielen, D. J., Weiss, M., and Patel, M. K. Long-term energy efficiency analysis requires solid energy statistics: The case of the German basic chemical industry. *Energy*, **2012**, *44* (1), 1094–1106.

[5] Schüth, F. Heterogene Katalyse. Schlüsseltechnologie der chemischen Industrie. *Chemie in unserer Zeit*, **2006**, *40* (2), 92–103.

[6] Shiju, N. R. and Guliants, V. V. Recent developments in catalysis using nanostructured materials. *Applied Catalysis A: General*, **2009**, *356* (1), 1–17.

[7] Grasselli, R. K. Fundamental Principles of Selective Heterogeneous Oxidation Catalysis. *Topics in Catalysis*, **2002**, *21* (1-3), 79–88.

[8] Ertl, G., Knoetzinger, H., and Weitkamp, J., Eds. *Handbook of Heterogeneous Catalysis*. Wiley-VCH Verlag GmbH, Weinheim, Germany, **1997**.

[9] *Evonik starts construction of second methionine complex in Singapore - Evonik Industries AG*, **2018**. https://corporate.evonik.com/en/media/press_releases/Pages/news-details.aspx?newsid=62899. Accessed 08/08/2018.

[10] Lin, M. M. Selective oxidation of propane to acrylic acid with molecular oxygen. *Applied Catalysis A: General*, **2001**, *207* (1-2), 1–16.

[11] Zhao, C. and Wachs, I. Selective oxidation of propylene over model supported V2O5 catalysts: Influence of surface vanadia coverage and oxide support. *Journal of Catalysis*, **2008**, *257* (1), 181–189.

[12] Centi, G., Cavani, F., and Trifirò, F. *Selective oxidation by heterogeneous catalysis*. Fundamental and applied catalysis. Kluwer Acad./Plenum Publ, New York, **2001**.

[13] Liu, L., Ye, X. P., and Bozell, J. J. A comparative review of petroleum-based and bio-based acrolein production. *ChemSusChem*, **2012**, *5* (7), 1162–1180.

[14] Katryniok, B., Paul, S., Bellière-Baca, V., Rey, P., and Dumeignil, F. Glycerol dehydration to acrolein in the context of new uses of glycerol. *Green Chem.*, **2010**, *12* (12), 2079.

[15] *The Propylene Gap: How Can It Be Filled? - American Chemical Society*, **2018**. https://www.acs.org/content/acs/en/pressroom/cutting-edge-chemistry/the-propylene-gap-how-can-it-be-filled.html. Accessed 08/08/2018.

[16] Grasselli, R. K. Advances and future trends in selective oxidation and ammoxidation catalysis. *Catalysis Today*, **1999**, *49* (1-3), 141–153.

[17] Ozkan, U. S. and Watson, R. B. The structure–function relationships in selective oxidation reactions over metal oxides. *Catalysis Today*, **2005**, *100* (1-2), 101–114.

[18] Kniep, B. L., Ressler, T., Rabis, A., Girgsdies, F., Baenitz, M., Steglich, F., and Schlögl, R. Rational design of nanostructured copper-zinc oxide catalysts for the steam reforming of methanol. *Angewandte Chemie (International ed. in English)*, **2004**, *43* (1), 112–115.

[19] Wachs, I. E. Recent conceptual advances in the catalysis science of mixed metal oxide catalytic materials. *Catalysis Today*, **2005**, *100* (1-2), 79–94.

[20] Weckhuysen, B. M. Snapshots of a working catalyst: possibilities and limitations of in situ spectroscopy in the field of heterogeneous catalysis. *Chem. Commun.*, **2002**, (2), 97–110.

[21] Weckhuysen, B. M. Determining the active site in a catalytic process: Operando spectroscopy is more than a buzzword. *Phys. Chem. Chem. Phys.*, **2003**, *5* (20), 4351.

[22] Holleman, A. F., Wiberg, E., and Wiberg, N. *Lehrbuch der anorganischen Chemie*. de Gruyter, Berlin, New York, **2007**.

[23] Anastas, P. and Eghbali, N. Green chemistry: principles and practice. *Chemical Society reviews*, **2010**, *39* (1), 301–312.

[24] Enthaler, S., Junge, K., and Beller, M. Sustainable metal catalysis with iron: from rust to a rising star? *Angewandte Chemie (International ed. in English)*, **2008**, *47* (18), 3317–3321.

[25] Al-Fatesh, A. S., Fakeeha, A. H., Ibrahim, A. A., Khan, W. U., Atia, H., Eckelt, R., and Chowdhury, B. Iron Oxide Supported on Al2O3 Catalyst for Methane Decomposition Reaction. Effect of MgO Additive and Calcination Temperature. *Jnl Chinese Chemical Soc*, **2016**, *63* (2), 205–212.

[26] Védrine, J. C., Coudurier, G., and Millet, J.-M. M. Molecular design of active sites in partial oxidation reactions on metallic oxides. *Catalysis Today*, **1997**, *33* (1-3), 3–13.

[27] Grzybowska-Świerkosz, B. Thirty years in selective oxidation on oxides: what have we learned? *Topics in Catalysis*, **2000**, *11-12* (1-4), 23–42.

[28] Hagen, J. *Industrial catalysis. A practical approach*. Wiley-VCH, Weinheim, **2006**.

[29] Weckhuysen, B. M. and Keller, D. E. Chemistry, spectroscopy and the role of supported vanadium oxides in heterogeneous catalysis. *Catalysis Today*, **2003**, *78* (1-4), 25–46.

[30] Fierro, J.L.G. *Metal Oxides: Chemistry and Applications*. Taylor & Francis Group, **2010**.

[31] Tüysüz, H., Lehmann, C. W., Bongard, H., Tesche, B., Schmidt, R., and Schüth, F. Direct imaging of surface topology and pore system of ordered mesoporous silica (MCM-41, SBA-15, and KIT-6) and nanocast metal oxides by high resolution scanning electron microscopy. *J. Am. Chem. Soc.*, **2008**, *130* (34), 11510–11517.

[32] Yang, C.-M., Zibrowius, B., Schmidt, W., and Schüth, F. Stepwise Removal of the Copolymer Template from Mesopores and Micropores in SBA-15. *Chem. Mater.*, **2004**, *16* (15), 2918–2925.

[33] Zhao, D. Triblock Copolymer Syntheses of Mesoporous Silica with Periodic 50 to 300 Angstrom Pores. *Science*, **1998**, *279* (5350), 548–552.

[34] Zhao, D., Huo, Q., Feng, J., Chmelka, B. F., and Stucky, G. D. Nonionic Triblock and Star Diblock Copolymer and Oligomeric Surfactant Syntheses of Highly Ordered,

Hydrothermally Stable, Mesoporous Silica Structures. *J. Am. Chem. Soc.*, **1998**, *120* (24), 6024–6036.

[35] Impéror-Clerc, M., Davidson, P., and Davidson, A. Existence of a Microporous Corona around the Mesopores of Silica-Based SBA-15 Materials Templated by Triblock Copolymers. *J. Am. Chem. Soc.*, **2000**, *122* (48), 11925–11933.

[36] Beck, J. S., Vartuli, J. C., Roth, W. J., Leonowicz, M. E., Kresge, C. T., Schmitt, K. D., Chu, C. T. W., Olson, D. H., and Sheppard, E. W. A new family of mesoporous molecular sieves prepared with liquid crystal templates. *J. Am. Chem. Soc.*, **1992**, *114* (27), 10834–10843.

[37] Chao, M.-C., Chang, C.-H., Lin, H.-P., Tang, C.-Y., and Lin, C.-Y. Morphological control on SBA-15 mesoporous silicas via a slow self-assembling rate. *J Mater Sci*, **2009**, *44* (24), 6453–6462.

[38] Corma, A. From Microporous to Mesoporous Molecular Sieve Materials and Their Use in Catalysis. *Chem. Rev.*, **1997**, *97* (6), 2373–2419.

[39] Cano, L. A., Cagnoli, M. V., Fellenz, N. A., Bengoa, J. F., Gallegos, N. G., Alvarez, A. M., and Marchetti, S. G. Fischer-Tropsch synthesis. Influence of the crystal size of iron active species on the activity and selectivity. *Applied Catalysis A: General*, **2010**, *379* (1-2), 105–110.

[40] Cheng, K., Ordomsky, V. V., Virginie, M., Legras, B., Chernavskii, P. A., Kazak, V. O., Cordier, C., Paul, S., Wang, Y., and Khodakov, A. Y. Support effects in high temperature Fischer-Tropsch synthesis on iron catalysts. *Applied Catalysis A: General*, **2014**, *488*, 66–77.

[41] Park, J.-Y., Lee, Y.-J., Khanna, P. K., Jun, K.-W., Bae, J. W., and Kim, Y. H. Alumina-supported iron oxide nanoparticles as Fischer–Tropsch catalysts. Effect of particle size of iron oxide. *Journal of Molecular Catalysis A: Chemical*, **2010**, *323* (1-2), 84–90.

[42] Torres Galvis, H. M., Koeken, A. C.J., Bitter, J. H., Davidian, T., Ruitenbeek, M., Dugulan, A. I., and Jong, K. P. d. Effect of precursor on the catalytic performance of supported iron catalysts for the Fischer–Tropsch synthesis of lower olefins. *Catalysis Today*, **2013**, *215*, 95–102.

[43] Oschatz, M., Lamme, W. S., Xie, J., Dugulan, A. I., and Jong, K. P. d. Ordered Mesoporous Materials as Supports for Stable Iron Catalysts in the Fischer-Tropsch Synthesis of Lower Olefins. *ChemCatChem*, **2016**, *8* (17), 2846–2852.

[44] Herranz, T., Rojas, S., Pérez-Alonso, F. J., Ojeda, M., Terreros, P., and Fierro, J.L.G. Carbon oxide hydrogenation over silica-supported iron-based catalysts. *Applied Catalysis A: General*, **2006**, *308*, 19–30.

[45] Sun, Y., Walspurger, S., Tessonnier, J.-P., Louis, B., and Sommer, J. Highly dispersed iron oxide nanoclusters supported on ordered mesoporous SBA-15: A very active catalyst for Friedel–Crafts alkylations. *Applied Catalysis A: General*, **2006**, *300* (1), 1–7.

[46] Jiang, Y., Lin, K., Zhang, Y., Liu, J., Li, G., Sun, J., and Xu, X. Fe-MCM-41 nanoparticles as versatile catalysts for phenol hydroxylation and for Friedel–Crafts alkylation. *Applied Catalysis A: General*, **2012**, *445-446*, 172–179.

[47] Li, B., Wu, K., Yuan, T., Han, C., Xu, J., and Pang, X. Synthesis, characterization and catalytic performance of high iron content mesoporous Fe-MCM-41. *Microporous and Mesoporous Materials*, **2012**, *151*, 277–281.

[48] Zhang, Q., Li, Y., An, D., and Wang, Y. Catalytic behavior and kinetic features of FeOx/SBA-15 catalyst for selective oxidation of methane by oxygen. *Applied Catalysis A: General*, **2009**, *356* (1), 103–111.

[49] He, J., Li, Y., An, D., Zhang, Q., and Wang, Y. Selective oxidation of methane to formaldehyde by oxygen over silica-supported iron catalysts. *Journal of Natural Gas Chemistry*, **2009**, *18* (3), 288–294.

[50] Arena, F., Gatti, G., Martra, G., Coluccia, S., Stievano, L., Spadaro, L., Famulari, P., and Parmaliana, A. Structure and reactivity in the selective oxidation of methane to formaldehyde of low-loaded FeOx/SiO2 catalysts. *Journal of Catalysis*, **2005**, *231* (2), 365–380.

[51] Koekkoek, A.J.J., Kim, W., Degirmenci, V., Xin, H., Ryoo, R., and Hensen, E.J.M. Catalytic performance of sheet-like Fe/ZSM-5 zeolites for the selective oxidation of benzene with nitrous oxide. *Journal of Catalysis*, **2013**, *299*, 81–89.

[52] Jia, J. One-step oxidation of benzene to phenol with nitrous oxide over Fe/MFI catalysts. *Journal of Catalysis*, **2004**, *221* (1), 119–126.

[53] Wong, S.-T., Lee, J.-F., Cheng, S., and Mou, C.-Y. In-situ study of MCM-41-supported iron oxide catalysts by XANES and EXAFS. *Applied Catalysis A: General*, **2000**, *198* (1-2), 115–126.

[54] Genz, N. S. *Eisenoxid geträgert auf SBA-15 als Katalysator für die selektive Oxidation von Propen*. Masterarbeit, Technische Universität Berlin, **2015**.

[55] Gabelica, Z., Charmot, A., Vataj, R., Soulimane, R., Barrault, J., and Valange, S. Thermal degradation of iron chelate complexes adsorbed on mesoporous silica and alumina. *J Therm Anal Calorim*, **2009**, *95* (2), 445–454.

[56] Tsoncheva, T., Rosenholm, J., Linden, M., Ivanova, L., and Minchev, C. Iron and copper oxide modified SBA-15 materials as catalysts in methanol decomposition. Effect of copolymer template removal. *Applied Catalysis A: General*, **2007**, *318*, 234–243.

[57] Zhang, H., Tang, C., Sun, C., Qi, L., Gao, F., Dong, L., and Chen, Y. Direct synthesis, characterization and catalytic performance of bimetallic Fe–Mo-SBA-15 materials in selective catalytic reduction of NO with NH3. *Microporous and Mesoporous Materials*, **2012**, *151*, 44–55.

[58] Ma, W., Kugler, E. L., Wright, J., and Dadyburjor, D. B. Mo–Fe Catalysts Supported on Activated Carbon for Synthesis of Liquid Fuels by the Fischer–Tropsch Process: Effect of Mo Addition on Reducibility, Activity, and Hydrocarbon Selectivity. *Energy Fuels*, **2006**, *20* (6), 2299–2307.

[59] Kharaji, A. G., Shariati, A., and Takassi, M. A. A Novel γ-Alumina Supported Fe-Mo Bimetallic Catalyst for Reverse Water Gas Shift Reaction. *Chinese Journal of Chemical Engineering*, **2013**, *21* (9), 1007–1014.

[60] Qin, S., Zhang, C., Xu, J., Yang, Y., Xiang, H., and Li, Y. Fe–Mo interactions and their influence on Fischer–Tropsch synthesis performance. *Applied Catalysis A: General*, **2011**, *392* (1-2), 118–126.

[61] Huang, T., Huang, W., Huang, J., and Ji, P. Methane reforming reaction with carbon dioxide over SBA-15 supported Ni–Mo bimetallic catalysts. *Fuel Processing Technology*, **2011**, *92* (10), 1868–1875.

[62] Maiti, G. C., Malessa, R., and Baerns, M. Studies on the reduction of the Fe2O3MoO3 system and its interaction with synthesis gas (CO+H2). *Thermochimica Acta*, **1984**, *80* (1), 11–21.

[63] Mars, P. and van Krevelen, D. W. Oxidations carried out by means of vanadium oxide catalysts. *Chemical Engineering Science*, **1954**, *3*, 41–59.

[64] Krylov, O. V. and Margolis, L. Y. Selectivity of Partial Hydrocarbon Oxidation. *International Reviews in Physical Chemistry*, **1983**, *3* (3), 305–333.

[65] Otsuka, K. and Wang, Y. Direct conversion of methane into oxygenates. *Applied Catalysis A: General*, **2001**, *222* (1-2), 145–161.

[66] Kobayashi, T., Guilhaume, N., Miki, J., Kitamura, N., and Haruta, M. Oxidation of methane to formaldehyde over FeSiO2 and Sn W mixed oxides. *Catalysis Today*, **1996**, *32* (1-4), 171–175.

[67] Kobayashi, T. Selective oxidation of light alkanes to aldehydes over silica catalysts supporting mononuclear active sites — acrolein formation from ethane. *Catalysis Today*, **2001**, *71* (1-2), 69–76.

[68] Sing, K. S. W. Reporting physisorption data for gas/solid systems with special reference to the determination of surface area and porosity (Recommendations 1984). *Pure and Applied Chemistry*, **1985**, *57* (4), 603–619.

[69] Brunauer, S., Emmett, P. H., and Teller, E. Adsorption of Gases in Multimolecular Layers. *J. Am. Chem. Soc.*, **1938**, *60* (2), 309–319.

[70] Barrett, E. P., Joyner, L. G., and Halenda, P. P. The Determination of Pore Volume and Area Distributions in Porous Substances. I. Computations from Nitrogen Isotherms. *J. Am. Chem. Soc.*, **1951**, *73* (1), 373–380.

[71] Smith, M. A. and Lobo, R. F. A fractal description of pore structure in block-copolymer templated mesoporous silicates. *Microporous and Mesoporous Materials*, **2010**, *131* (1-3), 204–209.

[72] Pfeifer, P. Multilayer adsorption on a fractally rough surface. *Phys. Rev. Lett.*, **1989**, *62* (17), 1997–2000.

[73] Weckhuysen, B. M. *In-situ spectroscopy of catalysts*. American Scientific Publishers, Stevenson Ranch, Calif., **2004**.

[74] Dieterle, M., Weinberg, G., and Mestl, G. Raman spectroscopy of molybdenum oxides. *Phys. Chem. Chem. Phys.*, **2002**, *4* (5), 812–821.

[75] Kortüm, G. *Reflexionsspektroskopie. Grundlagen, Methodik, Anwendungen*. Springer-Verlag, Berlin, Heidelberg, New York, **1969**.

[76] Kubelka, P. and Munk, F. An article on optics of paint layers. *Zeitschrift für technische Physik*, **1931**, (12), 593–601.

[77] Niemantsverdriet, J. W. *Spectroscopy in catalysis. An introduction.* Wiley-VCH, Weinheim, Chichester, **2010**.

[78] Koningsberger, D. C., Ed. *X-ray absorption. Principles, applications, techniques of EXAFS, SEXAFS and XANES.* A Wiley Interscience publication, *92*. Wiley, New York, **1988**.

[79] Dinnebier, R. E. and Billinge, S. J. L. *Powder diffraction. Theory and practice.* Royal Society of Chemistry, Cambridge, **2008**.

[80] Spieß, L. *Moderne Röntgenbeugung. Röntgendiffraktometrie für Materialwissenschaftler, Physiker und Chemiker.* Studium. Teubner, Wiesbaden, **2008**.

[81] Zieliński, J., Zglinicka, I., Znak, L., and Kaszkur, Z. Reduction of Fe2O3 with hydrogen. *Applied Catalysis A: General*, **2010**, *381* (1-2), 191–196.

[82] Khawam, A. and Flanagan, D. R. Basics and applications of solid-state kinetics: a pharmaceutical perspective. *J Pharm Sci*, **2006**, *95* (3), 472–498.

[83] Vyazovkin, S. and Wight, C. A. Kinetics in solids. *Annu Rev Phys Chem*, **1997**, *48*, 125–149.

[84] Khawam, A. and Flanagan, D. R. Role of isoconversional methods in varying activation energies of solid-state kinetics. *Thermochimica Acta*, **2005**, *429* (1), 93–102.

[85] Vyazovkin, S. Kinetic concepts of thermally stimulated reactions in solids. A view from a historical perspective. *International Reviews in Physical Chemistry*, **2000**, *19* (1), 45–60.

[86] Genz, N. S., Baabe, D., and Ressler, T. Solid-State Kinetic Investigations of Nonisothermal Reduction of Iron Species Supported on SBA-15. *Journal of Analytical Methods in Chemistry*, **2017**, *2017* (1), 1–13.

[87] Kissinger, H. E. Reaction Kinetics in Differential Thermal Analysis. *Anal. Chem.*, **1957**, *29* (11), 1702–1706.

[88] Ozawa, T. Kinetic analysis of derivative curves in thermal analysis. *Journal of Thermal Analysis*, **1970**, *2* (3), 301–324.

[89] Flynn, J. H. The isoconversional method for determination of energy of activation at constant heating rates. *Journal of Thermal Analysis*, **1983**, *27* (1), 95–102.

[90] Gottwald, W. *GC für Anwender.* Die Praxis der instrumentellen Analytik. VCH, Weinheim, **1995**.

[91] Grob, R. L. *Modern Practice of Gas Chromatography (Fourth Edition).* s.n, s.l., **2004**.

[92] Cammann, K. *Instrumentelle analytische Chemie. Verfahren, Anwendungen und Qualitätssicherung.* Spektrum-Lehrbuch. Spektrum, Akad. Verl., Heidelberg, Berlin, **2001**.

[93] Hübschmann, H.-J. *Handbuch der GC/MS. Grundlagen und Anwendung.* VCH, Weinheim, **2005**.

[94] *DIN 51005. Thermische Analyse (TA)*, **2005**.

[95] Skoog, D. A. and Leary, J. J. *Instrumentelle Analytik. Grundlagen - Geräte - Anwendungen.* Springer-Lehrbuch. Springer, Berlin, **1996**.

[96] Haines, P. J. *Principles of thermal analysis and calorimetry.* RSC paperbacks. Royal Society of Chemistry, Cambridge, **2002**.

[97] Gütlich, P. Physikalische Methoden in der Chemie: Mößbauer-Spektroskopie I. *Chemie in unserer Zeit*, **1970**, *4* (5), 133–144.

[98] Gütlich, P. Physikalische Methoden in der Chemie: Mößbauer-Spektroskopie II. *Chemie in unserer Zeit*, **1971**, *5* (5), 131–141.

[99] Gauglitz, G. and Vo-Dinh, T., Eds. *Handbook of spectroscopy*. Wiley-VCH, Weinheim, **2005**.

[100] Zhao, D., Sun, J., Li, Q., and Stucky, G. D. Morphological Control of Highly Ordered Mesoporous Silica SBA-15. *Chem. Mater.*, **2000**, *12* (2), 275–279.

[101] Kruk, M., Jaroniec, M., Ko, C. H., and Ryoo, R. Characterization of the Porous Structure of SBA-15. *Chem. Mater.*, **2000**, *12* (7), 1961–1968.

[102] Zhuravlev, L. T. The surface chemistry of amorphous silica. Zhuravlev model. *Colloids and Surfaces A: Physicochemical and Engineering Aspects*, **2000**, *173* (1-3), 1–38.

[103] Blume, M. and Tjon, J. A. Mössbauer Spectra in a Fluctuating Environment. *Phys. Rev.*, **1968**, *165* (2), 446–456.

[104] Ressler, T. WinXAS: a program for X-ray absorption spectroscopy data analysis under MS-Windows. *Journal of synchrotron radiation*, **1998**, *5* (2), 118–122.

[105] Rehr, J. J., Booth, C. H., Bridges, F., and Zabinsky, S. I. X-ray-absorption fine structure in embedded atoms. *Phys. Rev. B*, **1994**, *49* (17), 12347–12350.

[106] Wachs, I. E. Raman and IR studies of surface metal oxide species on oxide supports. Supported metal oxide catalysts. *Catalysis Today*, **1996**, *27* (3-4), 437–455.

[107] Roozeboom, F., Fransen, T., Mars, P., and Gellings, P. J. Vanadium oxide monolayer catalysts. I. Preparation, characterization, and thermal stability. *Z. Anorg. Allg. Chem.*, **1979**, *449* (1), 25–40.

[108] Scholz, J., Walter, A., and Ressler, T. Influence of MgO-modified SBA-15 on the structure and catalytic activity of supported vanadium oxide catalysts. *Journal of Catalysis*, **2014**, *309*, 105–114.

[109] Gao, X. and Wachs, I. E. Investigation of Surface Structures of Supported Vanadium Oxide Catalysts by UV–vis–NIR Diffuse Reflectance Spectroscopy. *J Phys Chem B*, **2000**, *104* (6), 1261–1268.

[110] Koch, G., Schmack, L., and Ressler, T. Tuning Size and Reducibility of Copper Oxide Particles Supported on SBA-15. *ChemistrySelect*, **2016**, *1* (9), 2040–2049.

[111] Li, Y., Feng, Z., Lian, Y., Sun, K., Zhang, L., Jia, G., Yang, Q., and Li, C. Direct synthesis of highly ordered Fe-SBA-15 mesoporous materials under weak acidic conditions. *Microporous and Mesoporous Materials*, **2005**, *84* (1-3), 41–49.

[112] Arsene, D., Catrinescu, C., Dragoi, B., Teodosiu, C. Catalytic wet hydrogen peroxide oxidation of 4-chlorophenol over iron-exchanged clays. *Environmental Engineering and Management Journal*, **2010**, *9* (1), 7–16.

[113] Bordiga, S. Structure and Reactivity of Framework and Extraframework Iron in Fe-Silicalite as Investigated by Spectroscopic and Physicochemical Methods. *Journal of Catalysis*, **1996**, *158* (2), 486–501.

[114] Weber, R. S. Effect of Local Structure on the UV-Visible Absorption Edges of Molybdenum Oxide Clusters and Supported Molybdenum Oxides. *Journal of Catalysis*, **1995**, *151* (2), 470–474.

[115] Abe, T., Tachibana, Y., Uematsu, T., and Iwamoto, M. Preparation and characterization of Fe2O3 nanoparticles in mesoporous silicate. *J. Chem. Soc., Chem. Commun.*, **1995**, (16), 1617.

[116] Wang, Y., Yang, W., Yang, L., Wang, X., and Zhang, Q. Iron-containing heterogeneous catalysts for partial oxidation of methane and epoxidation of propylene. *Catalysis Today*, **2006**, *117* (1-3), 156–162.

[117] Kumar, M. S., Pérez-Ramírez, J., Debbagh, M. N., Smarsly, B., Bentrup, U., and Brückner, A. Evidence of the vital role of the pore network on various catalytic conversions of N2O over Fe-silicalite and Fe-SBA-15 with the same iron constitution. *Applied Catalysis B: Environmental*, **2006**, *62* (3-4), 244–254.

[118] He, Y. P. Size and structure effect on optical transitions of iron oxide nanocrystals. *Phys. Rev. B*, **2005**, *71* (12), 125411.

[119] Jitianu, A., Crisan, M., Meghea, A., Rau, I., and Zaharescu, M. Influence of the silica based matrix on the formation of iron oxide nanoparticles in the Fe2O3–SiO2 system, obtained by sol–gel method. *J. Mater. Chem.*, **2002**, *12* (5), 1401–1407.

[120] Faria, D. L. A. de, Venâncio Silva, S., and Oliveira, M. T. de. Raman microspectroscopy of some iron oxides and oxyhydroxides. *J. Raman Spectrosc.*, **1997**, *28* (11), 873–878.

[121] Nasibulin, A. G., Rackauskas, S., Jiang, H., Tian, Y., Mudimela, P. R., Shandakov, S. D., Nasibulina, L. I., Jani, S., and Kauppinen, E. I. Simple and rapid synthesis of α-Fe2O3 nanowires under ambient conditions. *Nano Res.*, **2009**, *2* (5), 373–379.

[122] Maxim, N., Overweg, A., Kooyman, P. J., van Wolput, J. H. M. C., Hanssen, R. W. J. M., van Santen, R. A., and Abbenhuis, H. C. L. Synthesis and Characterization of Microporous Fe–Si–O Materials with Tailored Iron Content from Silsesquioxane Precursors. *J Phys Chem B*, **2002**, *106* (9), 2203–2209.

[123] García-Aguilar, J., Cazorla-Amorós, D., and Berenguer-Murcia, Á. K- and Ca-promoted ferrosilicates for the gas-phase epoxidation of propylene with O 2. *Applied Catalysis A: General*, **2017**, *538*, 139–147.

[124] Li, Y., Feng, Z., Xin, H., Fan, F., Zhang, J., Magusin, P. C. M. M., Hensen, E. J. M., van Santen, R. A., Yang, Q., and Li, C. Effect of aluminum on the nature of the iron species in Fe-SBA-15. *J Phys Chem B*, **2006**, *110* (51), 26114–26121.

[125] Xiong, G., Li, C., Li, H., Xin, Q., and Feng, Z. Direct spectroscopic evidence for vanadium species in V-MCM-41 molecular sieve characterized by UV resonance Raman spectroscopy. *Chem. Commun.*, **2000**, (8), 677–678.

[126] Yang, S., Zhu, W., Zhang, Q., and Wang, Y. Iron-catalyzed propylene epoxidation by nitrous oxide: Effect of boron on structure and catalytic behavior of alkali metal ion-modified FeOx/SBA-15. *Journal of Catalysis*, **2008**, *254* (2), 251–262.

[127] Fan, F., Feng, Z., and Li, C. UV Raman spectroscopic studies on active sites and synthesis mechanisms of transition metal-containing microporous and mesoporous materials. *Accounts of chemical research*, **2010**, *43* (3), 378–387.

[128] Shebanova, O. N. and Lazor, P. Raman study of magnetite (Fe3O4): laser-induced thermal effects and oxidation. *J. Raman Spectrosc.*, **2003**, *34* (11), 845–852.

[129] Han, Q., Liu, Xu, Chen, Wang, and Zhang, H. Growth and Properties of Single-Crystalline γ-Fe2O3 Nanowires. *J. Phys. Chem. C*, **2007**, *111* (13), 5034–5038.

[130] Kündig, W., Bömmel, H., Constabaris, G., and Lindquist, R. H. Some Properties of Supported Small α–Fe2O3 Particles Determined with the Mössbauer Effect. *Phys. Rev.*, **1966**, *142* (2), 327–333.

[131] Oh, S. J., Cook, D. C., and Townsend, H. E. *Hyperfine Interactions*, **1998**, *112* (1/4), 59–66.

[132] Samanta, S., Giri, S., Sastry, P. U., Mal, N. K., Manna, A., and Bhaumik, A. Synthesis and Characterization of Iron-Rich Highly Ordered Mesoporous Fe-MCM-41. *Ind. Eng. Chem. Res.*, **2003**, *42* (13), 3012–3018.

[133] Boubnov, A., Lichtenberg, H., Mangold, S., and Grunwaldt, J.-D. Structure and reducibility of a Fe/Al 2 O 3 catalyst for selective catalytic reduction studied by Fe K-edge XAFS spectroscopy. *J. Phys.: Conf. Ser.*, **2013**, *430*, 12054.

[134] Farges, F., Lefrère, Y., Rossano, S., Berthereau, A., Calas, G., and Brown, G. E. The effect of redox state on the local structural environment of iron in silicate glasses: a combined XAFS spectroscopy, molecular dynamics, and bond valence study. *Journal of Non-Crystalline Solids*, **2004**, *344* (3), 176–188.

[135] Chaurand, P., Rose, J., Briois, V., Salome, M., Proux, O., Nassif, V., Olivi, L., Susini, J., Hazemann, J.-L., and Bottero, J.-Y. New methodological approach for the vanadium K-edge X-ray absorption near-edge structure interpretation: application to the speciation of vanadium in oxide phases from steel slag. *J Phys Chem B*, **2007**, *111* (19), 5101–5110.

[136] Vinu, A., Sawant, D. P., Ariga, K., Hossain, K. Z., Halligudi, S. B., Hartmann, M., and Nomura, M. Direct Synthesis of Well-Ordered and Unusually Reactive FeSBA-15 Mesoporous Molecular Sieves. *Chem. Mater.*, **2005**, *17* (21), 5339–5345.

[137] Petit, P.-E., Farges, F., Wilke, M., and Solé, V. A. Determination of the iron oxidation state in Earth materials using XANES pre-edge information. *Journal of synchrotron radiation*, **2001**, *8* (2), 952–954.

[138] Maslen, E. N., Streltsov, V. A., Streltsova, N. R., and Ishizawa, N. Synchrotron X-ray study of the electron density in α-Fe2O3. *Acta Crystallogr B Struct Sci*, **1994**, *50* (4), 435–441.

[139] Cornell, R. M. and Schwertmann, U. *The iron oxides. Structure, properties, reactions, occurrence and uses.* VCH, Weinheim, New York, Basel, Cambridge, Tokyo, **1996**.

[140] Mackay, A. L. β-Ferric Oxyhydroxide. *Mineral. mag. j. Mineral. Soc.*, **1960**, *32* (250), 545–557.

[141] Behrens, M. and Schlögl, R. How to Prepare a Good Cu/ZnO Catalyst or the Role of Solid State Chemistry for the Synthesis of Nanostructured Catalysts. *Z. Anorg. Allg. Chem.*, **2013**, *639* (15), 2683–2695.

[142] Nishida, K., Atake, I., Li, D., Shishido, T., Oumi, Y., Sano, T., and Takehira, K. Effects of noble metal-doping on Cu/ZnO/Al2O3 catalysts for water–gas shift reaction. *Applied Catalysis A: General*, **2008**, *337* (1), 48–57.

[143] Takehira, K. and Shishido, T. Preparation of supported metal catalysts starting from hydrotalcites as the precursors and their improvements by adopting "memory effect". *Catal Surv Asia*, **2007**, *11* (1-2), 1–30.

[144] Antonić-Jelić, T., Bosnar, S., Bronić, J., Subotić, B., and Škreblin, M. Experimental evidence of the "memory" effect of amorphous aluminosilicate gel precursors. *Microporous and Mesoporous Materials*, **2003**, *64* (1-3), 21–32.

[145] Audebrand, N., Auffrédic, J.-P., and Louër, D. X-ray Diffraction Study of the Early Stages of the Growth of Nanoscale Zinc Oxide Crystallites Obtained from Thermal Decomposition of Four Precursors. General Concepts on Precursor-Dependent Microstructural Properties. *Chem. Mater.*, **1998**, *10* (9), 2450–2461.

[146] Védrine, J. C. and Fechete, I. Heterogeneous partial oxidation catalysis on metal oxides. *Comptes Rendus Chimie*, **2016**, *19* (10), 1203–1225.

[147] Hossain, M. M. and Ahmed, S. Cu-based mixed metal oxide catalysts for WGSR: Reduction kinetics and catalytic activity. *Can. J. Chem. Eng.*, **2013**, *91* (8), 1450–1458.

[148] Ressler, T., Wienold, J., Jentoft, R.E., Timpe, O., and Neisius, T. Solid state kinetics of the oxidation of MoO2 investigated by time-resolved X-ray absorption spectroscopy. *Solid State Communications*, **2001**, *119* (3), 169–174.

[149] Khawam, A. and Flanagan, D. R. Solid-state kinetic models: basics and mathematical fundamentals. *J Phys Chem B*, **2006**, *110* (35), 17315–17328.

[150] Janković, B. Kinetic analysis of the nonisothermal decomposition of potassium metabisulfite using the model-fitting and isoconversional (model-free) methods. *Chemical Engineering Journal*, **2008**, *139* (1), 128–135.

[151] Doyle, C. D. Kinetic analysis of thermogravimetric data. *J. Appl. Polym. Sci.*, **1961**, *5* (15), 285–292.

[152] Luke, Y. L. and Abramowitz, M. *Mathematical functions and their approximations*. Academic Press, New York, **1975**.

[153] Senum, G. I. and Yang, R. T. Rational approximations of the integral of the Arrhenius function. *Journal of Thermal Anaysis*, **1977**, (11), 445–447.

[154] Heal, G. R. Evaluation of the integral of the Arrhenius function by a series of Chebyshev polynomials — use in the analysis of non-isothermal kinetics. *Thermochimica Acta*, **1999**, *340-341*, 69–76.

[155] Khachani, M., El Hamidi, A., Kacimi, M., Halim, M., and Arsalane, S. Kinetic approach of multi-step thermal decomposition processes of iron(III) phosphate dihydrate FePO4·2H2O. *Thermochimica Acta*, **2015**, *610*, 29–36.

[156] Chen, G., Shi, Z., Yu, J., Wang, Z., Xu, J., Gao, B., and Hu, X. Kinetic analysis of the non-isothermal decomposition of carbon monofluoride. *Thermochimica Acta*, **2014**, *589*, 63–69.

[157] Coats, A. W. and Redfern, J. P. Kinetic Parameters from Thermogravimetric Data. *Nature*, **1964**, *201* (4914), 68–69.

[158] Smith, A. J., Garciano, L. O., Tran, T., and Wainwright, M. S. Structure and Kinetics of Leaching for the Formation of Skeletal (Raney) Cobalt Catalysts. *Ind. Eng. Chem. Res.*, **2008**, *47* (5), 1409–1415.

[159] Gao, Z., Amasaki, I., and Nakada, M. A description of kinetics of thermal decomposition of calcium oxalate monohydrate by means of the accommodated Rn model. *Thermochimica Acta*, **2002**, *385* (1-2), 95–103.

[160] Matusita, K., Komatsu, T., and Yokota, R. Kinetics of non-isothermal crystallization process and activation energy for crystal growth in amorphous materials. *J Mater Sci*, **1984**, *19* (1), 291–296.

[161] Lorente, E. Kinetic study of the redox process for separating and storing hydrogen. Oxidation stage and ageing of solid. *International Journal of Hydrogen Energy*, **2008**, *33* (2), 615–626.

[162] Joraid, A. A., Alamri, S. N., and Abu-Sehly, A. A. Model-free method for analysis of non-isothermal kinetics of a bulk sample of selenium. *Journal of Non-Crystalline Solids*, **2008**, *354* (28), 3380–3387.

[163] Lu, W., Yan, B., and Huang, W.-h. Complex primary crystallization kinetics of amorphous Finemet alloy. *Journal of Non-Crystalline Solids*, **2005**, *351* (40-42), 3320–3324.

[164] Weisz, P. B. and Prater, C. D. Interpretation of Measurements in Experimental Catalysis. *Advances in Catalysis and Related Subjects. Volume VI.* Academic Press, **1954**, 143–196.

[165] Madon, R. J. and Boudart, M. Experimental criterion for the absence of artifacts in the measurement of rates of heterogeneous catalytic reactions. *Ind. Eng. Chem. Fund.*, **1982**, *21* (4), 438–447.

[166] Kumar, D. and Ali, A. Nanocrystalline K–CaO for the transesterification of a variety of feedstocks: Structure, kinetics and catalytic properties. *Biomass and Bioenergy*, **2012**, *46*, 459–468.

[167] Patel, A. and Singh, S. A green and sustainable approach for esterification of glycerol using 12-tungstophosphoric acid anchored to different supports: Kinetics and effect of support. *Fuel*, **2014**, *118*, 358–364.

[168] Bettahar, M., Costentin, G., Savary, L., and Lavalley, J. On the partial oxidation of propane and propylene on mixed metal oxide catalysts. *Applied Catalysis A: General*, **1996**, *145* (1-2), 1–48.

[169]   Haber, J. and Turek, W. Kinetic Studies as a Method to Differentiate between Oxygen Species Involved in the Oxidation of Propene. *Journal of Catalysis*, **2000**, *190* (2), 320–326.

[170]   Védrine, J. C. Revisiting active sites in heterogeneous catalysis: Their structure and their dynamic behaviour. *Applied Catalysis A: General*, **2014**, *474*, 40–50.

[171]   Wimmers, O. J., Arnoldy, P., and Moulijn, J. A. Determination of the reduction mechanism by temperature-programmed reduction. Application to small iron oxide (Fe2O3) particles. *J. Phys. Chem.*, **1986**, *90* (7), 1331–1337.

[172]   Scholz, J., Walter, A., Hahn, A.H.P., and Ressler, T. Molybdenum oxide supported on nanostructured MgO: Influence of the alkaline support properties on MoOx structure and catalytic behavior in selective oxidation. *Microporous and Mesoporous Materials*, **2013**, *180*, 130–140.

[173]   Grasselli, R. K. Genesis of site isolation and phase cooperation in selective oxidation catalysis. *Topics in Catalysis*, **2001**, *15* (2/4), 93–101.

[174]   Ungureanu, A., Dragoi, B., Chirieac, A., Royer, S., Duprez, D., and Dumitriu, E. Synthesis of highly thermostable copper-nickel nanoparticles confined in the channels of ordered mesoporous SBA-15 silica. *J. Mater. Chem.*, **2011**, *21* (33), 12529.

[175]   Sietsma, J. R. A., Meeldijk, J. D., Versluijs-Helder, M., Broersma, A., van Dillen, A. J., Jongh, P. E. d., and Jong, K. P. d. Ordered Mesoporous Silica to Study the Preparation of Ni/SiO2 ex Nitrate Catalysts: Impregnation, Drying, and Thermal Treatments. *Chem. Mater.*, **2008**, *20* (9), 2921–2931.

[176]   Huirache-Acuña, R., Pawelec, B., Rivera-Muñoz, E., Nava, R., Espino, J., and Fierro, J.L.G. Comparison of the morphology and HDS activity of ternary Co-Mo-W catalysts supported on P-modified SBA-15 and SBA-16 substrates. *Applied Catalysis B: Environmental*, **2009**, *92* (1-2), 168–184.

[177]   Lou, Y., Wang, H., Zhang, Q., and Wang, Y. SBA-15-supported molybdenum oxides as efficient catalysts for selective oxidation of ethane to formaldehyde and acetaldehyde by oxygen. *Journal of Catalysis*, **2007**, *247* (2), 245–255.

[178]   Santhosh Kumar, M., Schwidder, M., Grünert, W., Bentrup, U., and Brückner, A. Selective reduction of NO with Fe-ZSM-5 catalysts of low Fe content: Part II. Assessing the function of different Fe sites by spectroscopic in situ studies. *Journal of Catalysis*, **2006**, *239* (1), 173–186.

[179]   Tian, H., Wachs, I. E., and Briand, L. E. Comparison of UV and visible Raman spectroscopy of bulk metal molybdate and metal vanadate catalysts. *J Phys Chem B*, **2005**, *109* (49), 23491–23499.

[180]   Rashad, M. M., Ibrahim, A. A., Rayan, D. A., Sanad, M.M.S., and Helmy, I. M. Photo-Fenton-like degradation of Rhodamine B dye from waste water using iron molybdate catalyst under visible light irradiation. *Environmental Nanotechnology, Monitoring & Management*, **2017**, *8*, 175–186.

[181]   Kong, L., Li, J., Zhao, Z., Liu, Q., Sun, Q., Liu, J., and Wei, Y. Oxidative dehydrogenation of ethane to ethylene over Mo-incorporated mesoporous SBA-16 catalysts: The effect of MoOx dispersion. *Applied Catalysis A: General*, **2016**, *510*, 84–97.

[182]   Jander, G. and Winkel, A. Über amphotere Oxydhydrate, deren wäßrige Lösungen und Kristallisierende Verbindungen. XII Mitteilung. Hydrolysierende Systeme und ihre Aggregationsprodukte mit besonderer Berücksichtigung der Erscheinungen in wäßrigen Aluminiumsalzlösungen. *Z. Anorg. Allg. Chem.*, **1931**, *200* (1), 257–278.

[183]   Hu, H., Wachs, I. E., and Bare, S. R. Surface Structures of Supported Molybdenum Oxide Catalysts: Characterization by Raman and Mo L3-Edge XANES. *J. Phys. Chem.*, **1995**, *99* (27), 10897–10910.

[184]   Deo, G. and Wachs, I. E. Predicting molecular structures of surface metal oxide species on oxide supports under ambient conditions. *J. Phys. Chem.*, **1991**, *95* (15), 5889–5895.

[185]   Williams, C. C., Ekerdt, J. G., Jehng, J. M., Hardcastle, F. D., Turek, A. M., and Wachs, I. E. A Raman and ultraviolet diffuse reflectance spectroscopic investigation of silica-supported molybdenum oxide. *J. Phys. Chem.*, **1991**, *95* (22), 8781–8791.

[186]   Boulaoued, A., Fechete, I., Donnio, B., Bernard, M., Turek, P., and Garin, F. Mo/KIT-6, Fe/KIT-6 and Mo–Fe/KIT-6 as new types of heterogeneous catalysts for the conversion of MCP. *Microporous and Mesoporous Materials*, **2012**, *155*, 131–142.

[187]   Zhang, H., Shen, J., and Ge, X. The Reduction Behavior of Fe-Mo-O Catalysts Studied by Temperature-Programmed Reduction Combined with in Situ Mössbauer Spectroscopy and X-Ray Diffraction. *Journal of Solid State Chemistry*, **1995**, *117* (1), 127–135.

10 Figure captions

11 Table captions

Table 3.1: Characterization methods together with the provided information on supported iron oxide-based catalysts and support material. N_2 physisorption, diffuse reflectance UV-Vis spectroscopy (DR-UV-Vis), X-ray absorption spectroscopy (XAS), powder X-ray diffraction (XRD), temperature-programmed reduction (TPR), thermal analysis (TA), Mössbauer spectroscopy (MBS), Raman spectroscopy (Raman), and X-ray fluorescence spectroscopy (XRF). 11

Table 4.1: Structural properties of SBA-15 support material before and after catalytic reaction. Specific surface area, $a_{s,BET}$, pore volume, V_{pore}, average pore radius, r_p, lattice constant, a_0, corresponding to the hexagonal unit cell, and wall thickness, d_w, between the mesopores of SBA-15. Catalytic reaction was conducted in 5% propene and 5% oxygen in helium at 653 K. 29

Table 5.1: Surface and porosity characteristics of Fe_xO_y/SBA-15 samples. Specific surface area, $a_{s,BET}$, pore volume, V_{pore}, ratio of mesopore surface area, a_{pore}, and specific surface area, $a_{s,BET}$, as measure of micropore contribution to the entire surface of SBA-15, differences in fractal dimension, ΔD_f, between SBA-15 and corresponding Fe_xO_y/SBA-15 samples as measure of the roughness of the surface, and surface coverage, $\Phi_{Fe\ atoms}$. 41

Table 5.2: Lattice constant, a_0, and wall thickness, d_w, of Fe_xO_y/SBA-15 samples. 45

Table 5.3: Mössbauer parameters for 9.3wt% Fe_Nitrate, 7.2wt% Fe_Nitrate, 2.0wt% Fe_Nitrate, and 10.7wt% Fe_Citrate. Temperature, T, isomer shift, δ (referred to α-Fe at 298 K and not corrected for 2nd order Doppler shift), quadrupole shift, ε, line widths, Γ_{HWHM}, hyperfine magnetic field, B_{hf}, fluctuation rate, v_c, and area. [*] values held fixed in simulation. [a] relaxation rate reached the dynamic limit. 57

Table 5.4: Overview of results of applied spectroscopic methods for characterization of Fe_xO_y/SBA-15 samples. ✔: method corroborated result. ✘: no conclusion possible. 58

Table 5.5: Average iron oxidation state of Fe_xO_y/SBA-15 samples calculated from linear calibration (eq.(5.5)). 61

Table 5.6: Fraction of α-Fe_2O_3- and β-FeOOH-like species in Fe_xO_y/SBA-15 obtained from phase composition analysis of Fe K edge XANES spectra. Corresponding refinements are shown in Figure 5.21. 63

Table 5.7: EXAFS fit results for Fe_xO_y/SBA-15 samples and α-Fe_2O_3. Type of neighbor, coordination number, N, and XAFS disorder parameter, σ^2, of atoms at the distance R from the Fe atoms in Fe_xO_y/SBA-15 samples at various iron loadings. Experimental parameters were obtained from refinement of α-Fe_2O_3 model structure (AMCSD 0017806 [137]) to the experimental Fe K edge FT($\chi(k)*k^3$) of Fe_xO_y/SBA-15 (Figure 5.24). Fit range in R space 1.1-3.4 Å (or 1.1-3.0 Å); k space 2.2-10.3 $Å^{-1}$ (or 2.2-9.1 $Å^{-1}$); N_{ind} = 13.4 (or 11.3); N_{free} = 7 (or 5); Number of SS paths = 3 (or 2); S_0^2 = 0.9. Alternative values of fitting parameters given in parentheses refer to slightly modified refinement for sample 2.5wt% Fe_Citrate due to $\chi(k)*k^3$ data quality. 67

12 Appendix

Table A.12.1: Iron and molybdenum reference compounds together with their purchase information.

Reference compound	Purity and provider
α-Fe_2O_3	99.98%, Roth
Fe_3O_4	99.99%, Sigma Aldrich
FeO	99.8% trace metal basis, Sigma Aldrich
β-FeOOH	Catalyst grade, Sigma Aldrich
$FeC_2O_4 \cdot 2H_2O$	$\geq$ 98%, Roth
γ-Fe_2O_3	Nanopowder, Sigma Aldrich
$C_6H_8O_7 \cdot xFe^{3+} \cdot yNH_3$	~18% Fe, Roth
$Fe(NO_3)_3 \cdot 9H_2O$	99+% for analysis, ACROS Organics
MoO_3	99.5%, Sigma Aldrich
$(NH_4)_6Mo_7O_{24} \cdot 4H_2O$	99.0%, Fluka

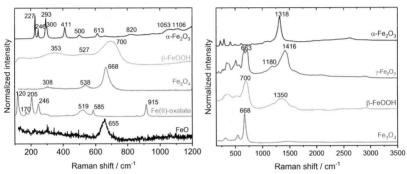

Figure A.12.1: Left: Raman spectra of various iron references. Right: Raman spectra of main iron oxide references measured in an extended Raman shift range.

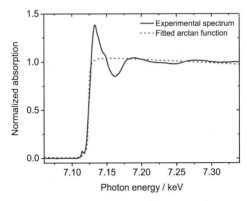

Figure A.12.2: Fitting routine to determine Fe K edge position of an experimental XANES spectrum (exemplary depicted for 7.2wt% Fe_Nitrate).

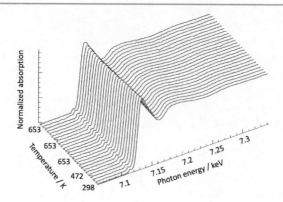

Figure A.12.3: *In situ* XANES spectra of 10.7wt% Fe_Citrate during selective oxidation of propene. (5% propene, 5% oxygen in helium, temperature range 298–653 K, heating rate 5 K/min).

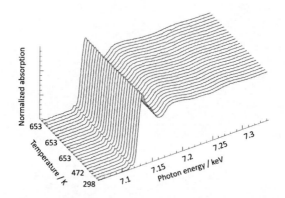

Figure A.12.4: *In situ* XANES spectra of 9.3wt% Fe_Nitrate during selective oxidation of propene. (5% propene, 5% oxygen in helium, temperature range 298–653 K, heating rate 5 K/min).

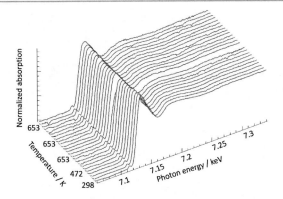

Figure A.12.5: *In situ* XANES spectra of 2.5wt% Fe_Citrate during selective oxidation of propene. (5% propene, 5% oxygen in helium, temperature range 298–653 K, heating rate 5 K/min).

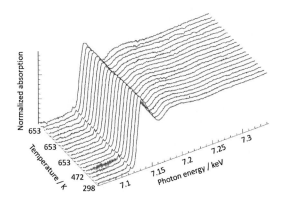

Figure A.12.6: *In situ* XANES spectra of 2.0wt% Fe_Nitrate during selective oxidation of propene. (5% propene, 5% oxygen in helium, temperature range 298–653 K, heating rate 5 K/min).

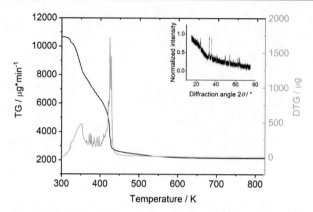

Figure A.12.7: Thermogram of Fe(III) nitrate nonahydrate during thermal treatment in 20% oxygen in helium. Temperature was increased with 5 K/min to 823 K. Left axis: TG curve. Right axis: DTG curve. Inset depicts XRD pattern of decomposition product corresponding to α-Fe_2O_3.

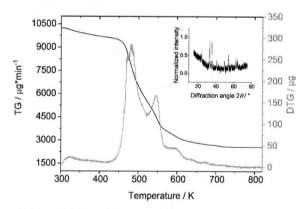

Figure A.12.8: Thermogram of (NH_4, Fe(III)) citrate during thermal treatment in 20% oxygen in helium. Temperature was increased with 5 K/min to 823 K. Left axis: TG curve. Right axis: DTG curve. Inset depicts XRD pattern of decomposition product corresponding to α-Fe_2O_3.

Epilogue

This work was performed between October 2015 and December 2018 at the Technical University Berlin under the guidance of Prof. Dr. Thorsten Ressler. Parts of this work have previously been published or presented at scientific conferences.

Publications

N. S. Genz, D. Baabe, T. Ressler, *Solid-state kinetic investigations of non-isothermal reduction of iron species supported on SBA-15*, Journal of Analytical Methods in Chemistry, **2017**, *2017* (1), 1–13.

Talks

N. S. Genz and T. Ressler, *Structure-activity correlations of Fe_xO_y/SBA-15 model catalysts for selective oxidation of propene*, 7. Berliner Chemie Symposium, April 5th 2018, Berlin, Germany.

N. S. Genz and T. Ressler, *Structure, kinetics, and reactivity of Fe_xO_y/SBA-15 selective oxidation catalysts*, Third Inorganic Chemistry Conference Erlangen, September 5th-8th 2017, Erlangen, Germany.

N. S. Genz and T. Ressler, *Influence of precursors on structural and kinetic properties of Fe_xO_y/SBA-15 model catalysts*, 6. Berliner Chemie Symposium, April 6th 2017, Berlin, Germany.

Posters

N. S. Genz and T. Ressler, *Structure and reactivity of iron oxide-based catalysts supported on nanostructured SBA-15*, 19. Vortragstagung für Anorganische Chemie der Fachgruppen Wöhler-Vereinigung und Festkörperchemie und Materialforschung, September 24th-27th 2018, Regensburg, Germany.

N. S. Genz and T. Ressler, *Non-isothermal reduction kinetics of Fe_xO_y/SBA-15 catalysts as starting point for structure-activity correlations*, 51. Jahrestreffen Deutscher Katalytiker, March 14th-16th 2018, Weimar, Germany.

N. S. Genz and T. Ressler, *Structural and kinetic properties of iron species supported on nanostructured SBA-15 as selective oxidation catalysts*, 13th European Congress on Catalysis, August 27th-31st 2017, Florence, Italy.

N. S. Genz and T. Ressler, *Structural and kinetic characterization of iron species supported on nanostructured SBA-15 as selective oxidation catalysts*, 50. Jahrestreffen Deutscher Katalytiker, March 15th-17th 2017, Weimar, Germany.

N. S. Genz and T. Ressler, *Iron oxides supported on nanostructured SBA-15 as model catalysts for selective oxidation of propene*, 49. Jahrestreffen Deutscher Katalytiker, March 16th-18th 2016, Weimar, Germany.

Danksagung

Mein besonderer Dank gilt Herrn Prof. Dr. Thorsten Ressler für die interessante wissenschaftliche Thematik meines Dissertationsthemas. Insbesondere möchte ich mich für die Möglichkeit des selbstständigen Erarbeitens meines Forschungsthemas, die stete Diskussionsbereitschaft und konstruktive Unterstützung während der gesamten Zeit meiner Forschungstätigkeit in seiner Arbeitsgruppe bedanken. Ich danke außerdem Herrn Prof. Dr. Malte Behrens für die die Anfertigung des Zweitgutachtens und Frau Prof. Dr. Maria Andrea Mroginski für die Übernahme des Vorsitzes des Promotionsausschusses.

Des Weiteren möchte ich mich bei der gesamten Arbeitsgruppe von Herrn Prof. Dr. Ressler für die angenehme Arbeitsatmosphäre bedanken. Herrn Dr. Alexander Müller danke ich für die stete Diskussionsbereitschaft bei jeglichen Themen der Festkörperkinetik. Bei Frau Semiha Schwarz bedanke ich mich für die technische Hilfestellung bei der SBA-15-Synthese.

Für die Unterstützung im Rahmen ihrer Bachelorarbeit danke ich Frau Anastasiya Knoch. Herrn Dr. Paul Sprenger möchte ich für die Durchführung der Raman-Spektroskopie danken und für seine große Diskussionsbereitschaft. Mein Dank gilt ferner Herrn Dr. Dirk Baabe für die Durchführung der Mössbauer-Spektroskopie und die unermüdliche Bereitschaft die bestmöglichen Ergebnisse zu erzielen. Für die Durchführung der Transmissionselektronen-mikroskopie-Messungen gilt mein Dank der Zentraleinrichtung Elektronenmikroskopie der TU Berlin. Der gesamten Arbeitsgruppe von Herrn Prof. Dr. Martin Lerch danke ich für die Durchführung der Weitwinkel-XRD-Messungen. Darüber hinaus möchte ich mich beim DESY in Hamburg für die Bereitstellung der Messzeit, sowie für die finanzielle Unterstützung bedanken. Des Weiteren danke ich dem Messzentrum der TU Berlin für die Durchführung der CHN-Analysen.

Darüber hinaus danke ich Herrn Dominik Knogler für die uneingeschränkte Unterstützung und den starken Rückhalt während meiner gesamten Forschungstätigkeit.